Methods in Molecular Biology

Series Editor:
John M. Walker
School of Life and Medical Sciences
University of Hertfordshire
Hatfield, Hertfordshire, AL10 9AB, UK

For further volumes:
http://www.springer.com/series/7651

Copy Number Variants

Methods and Protocols

Edited by

Derek M. Bickhart

Research Microbiologist/Bioinformatician, USDA ARS DFRC, Madison, WI, USA

Humana Press

Editor
Derek M. Bickhart
Research Microbiologist/Bioinformatician
USDA ARS DFRC
Madison, WI, USA

ISSN 1064-3745 ISSN 1940-6029 (electronic)
Methods in Molecular Biology
ISBN 978-1-4939-8665-1 ISBN 978-1-4939-8666-8 (eBook)
https://doi.org/10.1007/978-1-4939-8666-8

Library of Congress Control Number: 2018948178

Printed on acid-free paper

This Humana Press imprint is published by the registered company Springer Science+Business Media, LLC part of Springer Nature.
The registered company address is: 233 Spring Street, New York, NY 10013, U.S.A.

Preface

The detection of DNA copy number variants (CNVs) within the genomes of individuals has fascinated researchers since the foundation of modern genomics. CNVs represent large insertions, duplications, and deletions of DNA sequence in an individual's genome that range in size from 50 base pairs to millions of bases in size. These duplications and deletions segregate in the population, waxing and waning in frequency due to selective pressures or genetic drift. Their influence on the phenotype of the individual that harbors them can range from positive to deleterious; however, the majority of CNVs occur within the intergenic space of eukaryotic genomes and are therefore predicted to have neutral—or minor—effects.

When CNVs do overlap with gene regions, their larger sizes tend to encompass a majority of a gene, leading to speculation on the impact of the variation on downstream gene expression. While the impacts of CNVs on gene expression are often overestimated in the absence of experimental validation, the homozygous loss of a gene or gene regulatory region is highly likely to perturb gene expression networks. This makes CNV detection and classification an important consideration in genetics and pathway analysis.

Until recently, substantial computational expertise and statistical knowledge was required to operate most software designed for CNV detection. Bioinformaticians and computational biologists have needed to develop their own software to accurately detect CNVs in genetics data due to the novelty of the CNV variant type and the complexity of the data that was indicative of a CNV event. The cutting-edge nature of these analyses and the "rough edges" of open-source CNV detection software often restricted nonexperts from using them in their analysis workflows. Thankfully, the field has matured and CNV analysis and detection software has reached a critical juncture.

It is due to the recent development and constant refinement of highly accurate CNV calling methods and software that we found a need for a set of detailed protocols for detecting CNVs within biological datasets. In this volume, we hope to provide detailed instructions to the reader that enable beginners and experts alike to (a) run appropriate CNV detection software on a dataset of choice and (b) discern between false positive noise and true positive CNV signals. With the increasing expansion of SNP genotype and DNA sequence datasets, there will be an ever-present need to fully characterize all detectable genetic variation—CNVs among them—within each individual sample.

Madison, WI, USA *Derek M. Bickhart*

Contents

The original version of this book was revised. A correction to this book can be found at https://doi.org/10.1007/978-1-4939-8666-8_17

Contributors

CAN ALKAN • *Department of Computer Engineering, Bilkent University, Ankara, Turkey*
DEREK M. BICKHART • *Research Microbiologist/Bioinformatician, USDA ARS DFRC, Madison, WI, USA*
SVEN BOCKLANDT • *Bionano Genomics, San Diego, CA, USA*
MARK BORODKIN • *Bionano Genomics, San Diego, CA, USA*
OLA BRYNILDSRUD • *Norwegian Institute of Public Health, Oslo, Norway*
HAN CAO • *Bionano Genomics, San Diego, CA, USA*
SAKI CHAN • *Bionano Genomics, San Diego, CA, USA*
KEN CHEN • *Department of Bioinformatics and Computational Biology, The University of Texas MD Anderson Cancer Center, Houston, TX, USA*
RUOYAN CHEN • *Department of Paediatrics and Adolescent Medicine, LKS Faculty of Medicine, The University of Hong Kong, Pokfulam, Hong Kong*
ZECHEN CHONG • *Department of Genetics and Informatics Institute, School of Medicine, The University of Alabama at Birmingham, Birmingham, AL, USA*
LI FANG • *Raymond G. Perelman Center for Cellular and Molecular Therapeutics, Children's Hospital of Philadelphia, Philadelphia, PA, USA*
NATHAN FORTIER • *Golden Helix Inc., Bozeman, MT, USA*
LI GUO • *MOE Key Lab for Intelligent Networks & Network Security, Xi'an Jiaotong University, Xi'an, China*
KE HAO • *Department of Genetics and Genomic Sciences, Icahn Institute for Genomics and Multiscale Biology, Icahn School of Medicine at Mount Sinai, New York, NY, USA*
ALEX HASTIE • *Bionano Genomics, San Diego, CA, USA*
ERIK HOLMLIN • *Bionano Genomics, San Diego, CA, USA*
SERGII IVAKHNO • *Illumina Cambridge Ltd., Chesterford Research Park, Essex, UK*
FATMA KAHVECI • *Department of Computer Engineering, Bilkent University, Ankara, Turkey*
ERNEST LAM • *Bionano Genomics, San Diego, CA, USA*
ERIC-WUBBO LAMIJER • *MOE Key Lab for Intelligent Networks & Network Security, Xi'an Jiaotong University, Xi'an, China*
YU LUNG LAU • *Department of Paediatrics and Adolescent Medicine, LKS Faculty of Medicine, The University of Hong Kong, Pokfulam, Hong Kong*
JUNYA LI • *Institute of Animal Science, Beijing, China*
GEORGE E. LIU • *Animal Genomics and Improvement Laboratory, USDA ARS, Beltsville, MD, USA*
ZEMIN NING • *Wellcome Trust Sanger Institute, Hinxton, UK*
KEIRAN RAINE • *Wellcome Trust Sanger Institute, Hinxton, UK*
ERIC ROLLER • *Illumina Inc., San Diego, CA, USA*
GABE RUDY • *Golden Helix Inc., Bozeman, MT, USA*
MICHAEL SAGHBINI • *Bionano Genomics, San Diego, CA, USA*
ANDREAS SCHERER • *Golden Helix Inc., Bozeman, MT, USA*

ALEXANDER SCHLIEP • *Chalmers University of Technology, Gothenburg, Sweden*
KAI WANG • *Raymond G. Perelman Center for Cellular and Molecular Therapeutics, Children's Hospital of Philadelphia, Philadelphia, PA, USA*
JOHN WIEDENHOEFT • *Chalmers University of Technology, Gothenburg, Sweden; Rutgers University, New Brunswick, NJ, USA*
LINGYANG XU • *Institute of Animal Science, Beijing, China*
LIU YANG • *Institute of Animal Science, Beijing, China*
WANLING YANG • *Department of Paediatrics and Adolescent Medicine, LKS Faculty of Medicine, The University of Hong Kong, Pokfulam, Hong Kong*
XIAOFEI YANG • *MOE Key Lab for Intelligent Networks & Network Security, Xi'an Jiaotong University, Xi'an, China*
KAI YE • *MOE Key Lab for Intelligent Networks & Network Security, Xi'an Jiaotong University, Xi'an, China; School of Electronics and Information Engineering, Xi'an Jiaotong University, Xi'an, China; The First Affiliated Hospital of Xi'an Jiaotong University, Xi'an, China; Collaborative Innovation Center for Genetics and Development, School of Life Sciences, Fudan University, Shanghai, China*
ZHONGYANG ZHANG • *Department of Genetics and Genomic Sciences, Icahn Institute for Genomics and Multiscale Biology, Icahn School of Medicine at Mount Sinai, New York, NY, USA*
XUEFANG ZHAO • *Center for Genomic Medicine at Massachusetts General Hospital, Boston, MA, USA*
YANG ZHOU • *Huazhong Agricultural University, Wuhan, Hubei, China; Key Laboratory of Agricultural Animal Genetics, Breeding and Reproduction, Education Ministry of China, Huazhong Agricultural University, Wuhan, Hubei, China*

Chapter 1

Identification of Copy Number Variants from SNP Arrays Using PennCNV

Li Fang and Kai Wang

Abstract

High-resolution single-nucleotide polymorphism (SNP) genotyping arrays offer a sensitive and affordable method for genome-wide detection of copy number variants (CNVs). PennCNV is a hidden Markov model (HMM)-based CNV caller for SNP arrays, first released 10 years ago. A typical CNV calling procedure using PennCNV includes preparation of input files, CNV calling, filtering CNV calls, CNV annotation, and CNV visualization. Here we describe several protocols for CNV calling using PennCNV, together with descriptions on several recent improvements to the software tool.

Key words Copy number variants, SNP array, Hidden Markov model, PennCNV

1 Introduction

Copy number variants (CNVs) are DNA segments that are present at a variable copy number in comparison with a reference genome [1]. CNVs are a major source of genome diversity in human populations [2–4] and have been implicated in a variety of human diseases [5–8] and cancers [9, 10]. Microarray-based platforms have been developed for CNV detection [11, 12]. One of the major types of microarrays used for CNV detection is microarray Comparative Genomic Hybridization (array-CGH). Common CNV calling algorithms for array-CGH includes circular binary segmentation [13, 14], SW-ARRAY [15], among others. However, array-CGH is limited to detection of large CNVs that are tens or hundreds of kilobases [11]. Owing to the improved resolution and the ability to incorporate information from SNP alleles, high-resolution SNP genotyping arrays potentially offer a more sensitive method for genome-wide CNV detection [16].

There are two important measures of SNP signal intensities, including log R Ratio (LRR) and B Allele Frequency (BAF), in typical SNP arrays produced by Illumina (San Diego, CA). LRR is a measure of normalized total signal intensity, while BAF is a

Derek M. Bickhart (ed.), *Copy Number Variants: Methods and Protocols*, Methods in Molecular Biology, vol. 1833, https://doi.org/10.1007/978-1-4939-8666-8_1,

measure of normalized allelic intensity ratio [17]. The combination of LRR and BAF can be used together to determine different copy numbers and to differentiate copy-neutral LOH (loss of heterozygosity) regions from normal copy regions. For example, SNPs in a normal copy DNA segment in a diploid genome has three possible BAF values (0.0, 0.5, 1.0) and with LRR values centered around zero. In comparison, SNPs in a duplication region with three copies have four possible BAF values (0.0, 0.33, 0.67, 1.0) with increased LRR values.

Hidden Markov models (HMM) have been successfully applied to identify CNVs from SNP genotyping arrays. PennCNV [18] and QuantiSNP [19] are two widely used HMM-based CNV callers for Illumina SNP arrays. Both PennCNV and QuantiSNP incorporate LRR and BAF into the HMM model. In addition, PennCNV incorporates more information, including population allele frequency of each SNP and the distance between adjacent SNPs. PennCNV can also use family information when available, through either the trio-calling or the joint-calling algorithms. PennCNV was shown to be among the most reliable CNV callers in studies comparing the performance of different CNV callers on different SNP array platforms [20–22]. Because of its performance, PennCNV has been applied in a number of large-scale genetic studies [23–28]. In addition, some CNV postprocessing and association analysis tools (e.g., ParseCNV [29]) also use PennCNV as the default CNV caller. In this chapter, we describe the procedures to detect CNVs from SNP arrays using PennCNV. Currently it can handle signal intensity data from two major SNP array platforms: Illumina and Affymetrix. With appropriate preparation of file format, it can also handle other types of SNP arrays and oligonucleotide arrays (*see* **Note 3**).

PennCNV defines six hidden states, each corresponding to a different copy number state. In addition, PennCNV incorporates several components together to infer the hidden states, including LRR, BAF, the distance between neighboring SNPs, and the population frequency of the B allele. Detailed relationships between hidden states, copy numbers, CNV genotypes, BAF values are shown in Table 1.

A typical CNV calling procedure using PennCNV includes preparation of input files, CNV calling, filtering CNV calls and CNV annotation. PennCNV also provides functionality to visualize the CNV calls. The summary of PennCNV analysis pipeline is shown in Fig. 1. PennCNV requires signal intensity files (one file per sample), an HMM file, a PFB (Population frequency of B allele) file, and optionally a GCModel file as input files. Users of Illumina arrays can directly export LRR and BAF values from the GenomeStudio/BeadStudio software provided by Illumina. Users of Affymetrix arrays can prepare signal intensity files from raw CEL files using Affymetrix Power Tools software and the PennCNV-Affy

Table 1
Hidden states, copy numbers, CNV genotypes, and their descriptions

Copy number state	Total copy number	Description	CNV genotypes	BAF values
1	0	Deletion of two copies	Null	–
2	1	Deletion of one copy	A, B	0, 1
3	2	Normal state	AA, AB, BB	0, 0.5, 1
4	2	Copy-neutral with LOH	AA, BB	0, 1
5	3	Single copy duplication	AAA, AAB, ABB, BBB	0, 0.33, 0.67, 1
6	4	Double copy duplication	AAAA, AAAB, AABB, ABBB, BBBB	0, 0.25, 0.5, 0.75, 1

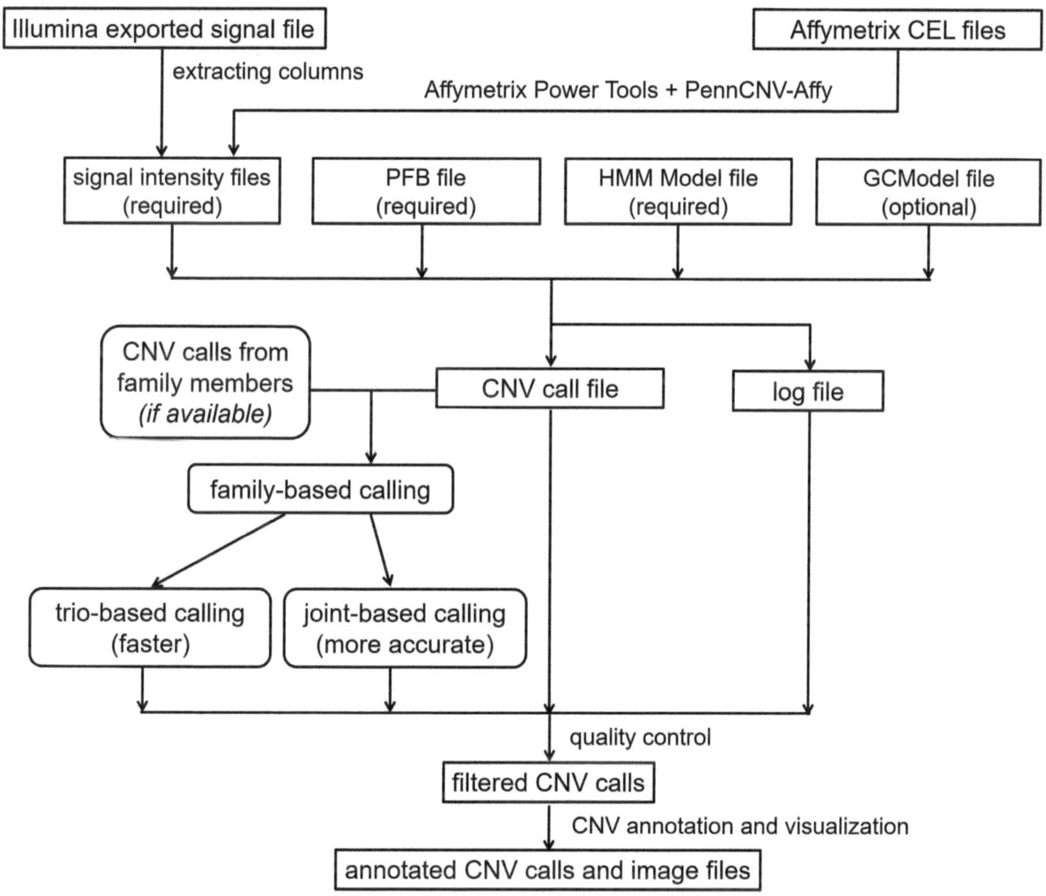

Fig. 1 Summary of PennCNV analysis pipeline

package. The CNV calling process will generate a raw CNV call file and a log file. If CNV calls from family members are available, family-based CNV calling (trio-based or joint-based) can be performed. Raw CNV calls can be further filtered to generate filtered CNV calls, which represent a call set with higher quality than the raw call set. In addition, PennCNV can perform functional annotation of the CNV calls to infer what genes, exons or genomic elements are disrupted by CNV calls, and can generate image files for signal intensity values around CNV calls for visualization and manual examination of calls to evaluate their reliability.

2 Materials

2.1 Equipment

A computer with internet connection. PennCNV can run on Windows system, but Linux or Unix-like operating systems are preferred.

2.2 Software Requirements

A C compiler (such as GCC) and a Perl interpreter are required to compile and run the PennCNV software. R is required to generate JPG/PDF files for signal intensity plots on CNV regions.

3 Methods

3.1 Installation of PennCNV

The PennCNV is available on GitHub. PennCNV is written in a mixture of Perl and C. In a typical Linux system, we can use the following commands to download and compile PennCNV:

```
git clone https://github.com/WGLab/PennCNV.git
cd PennCNV/kext
make
```

If no error occurs, the messages shown on the screen will be similar to below:

```
gcc 'perl -MExtUtils::Embed -e ccopts' -fPIC   -c -o
khmm_wrap.o khmm_wrap.c
gcc 'perl -MExtUtils::Embed -e ccopts' -fPIC   -c -o
khmm.o khmm.c
gcc 'perl -MExtUtils::Embed -e ccopts' -fPIC   -c -o
kc.o kc.c
gcc 'perl -MExtUtils::Embed -e ccopts' -fPIC   -c -o
khmmDev.o khmmDev.c
gcc -shared -o khmm.so khmm_wrap.o khmm.o kc.o
khmmDev.o 'perl -MExtUtils::Embed -e ldopts'
mkdir -p 'perl -MConfig -e 'print $Config{version}"
mkdir -p 'perl -MConfig -e 'print
$Config{version}"/'perl -MConfig -e 'print
$Config{archname}"/
```

```
mkdir -p 'perl -MConfig -e 'print
$Config{version}"/'perl -MConfig -e 'print
$Config{archname}"/auto/
mv khmm.so 'perl -MConfig -e 'print
$Config{version}"/'perl -MConfig -e 'print
$Config{archname}"/
```

After compiling PennCNV, we can go into the PennCNV/ directory and run the command: perl detect_cnv.pl. It will show the program usage information, indicating the successful installation of the program (*see* **Notes 1** and **2**).

3.2 Preparation of Input Files

PennCNV input files are all in text formats. It requires signal intensity files, an HMM file, a PFB (Population frequency of B allele) file, and optionally a GCModel file. Usually, users only need to prepare the signal intensity files and can use the default HMM file (hhall.hmm). For some commonly used SNP arrays, the default PFB and GCModel files may be downloaded from the PennCNV website (http://penncnv.openbioinformatics.org/en/latest/), but for other arrays, users need to generate these files by themselves. Below we describe these input file formats, and describe the procedure to prepare them.

3.2.1 Preparation of Signal Intensity Files

The input signal intensity file is a text file that contains information for one marker per line, and all fields in each line are tab-delimited. One example of the file is shown in Table 2.

The first line of the file specifies the meaning for each tab-delimited column. For example, there are six fields in each line in the file, corresponding to SNP name, chromosome, Position, genotype, LRR and BAF, respectively. The CNV calling only requires three fields: SNP name, LRR, and BAF. Genome coordinates of SNPs (chromosome and positions) are not required by default since PennCNV will read this information from the PFB file, as described later.

Table 2
An example of signal intensity file

Name	Chr	Position	99HI0698C.GType	99HI0698C.Log R Ratio	99HI0698C.B Allele Freq
rs13072188	3	38411	AA	0.1042794	0
rs9681213	3	41894	BB	0.07361082	0.9804617
rs1516321	3	57010	AA	0.06956207	0.01255646
rs1400176	3	70973	BB	−0.2123737	0.9924203

Users of Illumina SNP arrays usually have signal intensity files that contain LRR and BAF values. If not, LRR and BAF values can be extracted from Illumina report file following the online instructions (http://penncnv.openbioinformatics.org/en/latest/user-guide/input/).

Users of Affymetrix arrays can prepare signal intensity files from raw CEL files using Affymetrix Power Tools and the PennCNV-Affy package, following the instructions below. If you do not use Affymetrix arrays, you can skip this section and go to Subheading 3.2.2. The Penn-Affy workflow can be also adapted to other SNP array platforms (*see* **Note 3**).

PennCNV-Affy package supports several Affymetrix SNP arrays, including Genome-Wide Human SNP Array 6.0 [30], Genome-Wide Human SNP Array 5.0 and Human Mapping 500K Array Set, as well as other more recently developed arrays such as the various versions of Axiom arrays. Next, we will introduce the procedures to prepare signal intensity files from CEL files of Genome-Wide Human SNP Array 6.0. The example data set we used is freely available at NCBI GEO database (Accession No. GSE15826). The instructions on downloading this data set can be found in **Note 4**.

Step 1. Download software tools and libraries.

The Affymetrix Power Tools can be downloaded from the following web page:

https://www.thermofisher.com/us/en/home/life-science/microarray-analysis/microarray-analysis-partners-programs/affymetrix-developers-network/affymetrix-power-tools.html. Unzip the file, we will see a bin/ directory. We can add this directory to the PATH environmental variable so that the binary program files in the bin/ directory can be executed directly by typing the program name.

PennCNV-Affy package with libraries is provided within the PennCNV package. In the PennCNV/ directory, we can see an affy/ directory. Inside the affy/ directory, there are a few subdirectories that contain PennCNV-specific library files for various Affymetrix arrays.

In addition to the PennCNV-specific library files, we need to download another file used by Affymetrix Power Tools for this specific array. To download this file, we need to log in but the registration is free. The URL is:

http://www.affymetrix.com/Auth/support/downloads/library_files/genomewidesnp6_libraryfile.zip

We can use the following command to unzip the downloaded file:

```
unzip genomewidesnp6_libraryfile.zip
```

After unzipping the file, we will see a CD_GenomeWideSNP_6_rev3/ directory. The Affymetrix Power Tools library files we need are in the CD_GenomeWideSNP_6_rev3/Full/GenomeWideSNP_6/LibFiles/ directory. We can copy all the files in this directory to the PennCNV/affy/libgw6/directory using the following command:

```
cp CD_GenomeWideSNP_6_rev3/Full/GenomeWideSNP_6/
LibFiles/* PennCNV/affy/libgw6/
```

Step 2. Prepare CEL list file

Since we have a lot of input CEL files, we need to store the file names in a list file. The list file should contain one file name per line, with the first line being "cel_files". Assuming the raw CEL files are stored in the raw_data/directory, we can use the following commands to generate the input list file:

```
echo cel_files > input_cel_list
ls raw_data/*.CEL  >> input_cel_list
```

Step 3. Generate genotyping calls from CEL files

After downloading the software and library files, we can generate genotyping calls from CEL files using the following command:

```
apt-probeset-genotype --cdf-file PennCNV/affy/lib-
gw6/GenomeWideSNP_6.cdf --analysis birdseed --read-
models-birdseed PennCNV/affy/libgw6/GenomeWideSNP_6.
birdseed.models --special-snps PennCNV/affy/libgw6/
GenomeWideSNP_6.specialSNPs --cel-files input_cel_list
--out-dir output/
```

The above command generates genotyping calls using the Birdseed algorithm [31]. It performs a multiple-chip analysis to estimate signal intensity for each allele of each SNP, fitting probe-specific effects to increase precision. Four output files will be generated: birdseed.confidences.txt, birdseed.report.txt, birdseed.calls.txt and apt-probeset-genotype.log. The birdseed.calls.txt file contains the genotyping calls of each SNP and the birdseed.confidences.txt file contains the associated confidences. The birdseed.report.txt file contains some summary statistics of each SNP array. The log information is stored in apt-probeset-genotype.log.

Step 4. Extract the allele-specific signals

For each SNP, we have a signal measure for the A allele and a separate signal measure for the B allele. After generating genotyping calls, we can extract the allele-specific signals using the following command:

```
apt-probeset-summarize --cdf-file PennCNV/affy/
libgw6/GenomeWideSNP_6.cdf --analysis quant-norm.
sketch=50000,pm-only,med-polish,expr.genotype=true
--target-sketch PennCNV/affy/libgw6/hapmap.quant-norm.
normalization-target.txt --cel-files input_cel_list
--out-dir output/
```

The above command reads signal intensity values in the CEL files specified in input_cel_list, apply quantile normalization to the values, applies median polish on the data, then generates signal intensity values for A and B allele for each SNP. The file hapmap.quant-norm.normalization-target.txt is provided in the PennCNV-Affy package. It is generated using all HapMap samples, as a reference quantile distribution to use in the normalization process, so that the quantile normalization procedures for different genotyping projects are more comparable to each other. Three output files will be generated in the output/ directory, including quant-norm.pm-only.med-polish.expr.summary.txt, quant-norm.pm-only.med-polish.expr.report.txt, and apt-probeset-summarize.log. The quant-norm.pm-only.med-polish.expr.summary.txt file contains the signal values for A and B allele (*see* **Note 5**).

Step 5. Generate the canonical genotype clustering file

Next, we can use the generate_affy_geno_cluster.pl program in the PennCNV-Affy package to generate the canonical genotype clustering file. This file contains the parameters to calculate LRR and BAF values. The command is shown below.

```
perl PennCNV/affy/bin/generate_affy_geno_cluster.pl
output/birdseed.calls.txt output/birdseed.confidences.
txt output/quant-norm.pm-only.med-polish.expr.summary.
txt -locfile PennCNV/affy/libgw6/affygw6.hg38.pfb  -out
output/gw6.genocluster -sexfile cel_sex_file
```

The affygw6.hg38.pfb file is provided in PennCNV-Affy package, which contains the annotated marker positions in hg38 genome assembly. Detailed description of the file format is in Subheading 3.2.3. The cel_sex_file is a two-column file that annotates the sex information for each CEL file, one file per line, and each line contains the file name and the sex (separated by tab).The cel_sex_file is important for chrX markers and chrY markers (*see* **Notes 6** and 7)

The output file (gw6.genocluster) is a tab-delimited text file. The first a few lines of this file is shown in Table 3. It contains 10 columns. The first line of the file specifies the meaning for each column. The first column is the probe set ID (marker ID); columns 2–4 are the R values for the three canonical genotypes (AA, AB, BB); columns 5–7 are the θ values for the three canonical genotypes; columns 8–10 are number of arrays of each genotype. R and θ values are parameters for calculating LRR and BAF values [32].

Step 6. Calculate the LRR and BAF values

Next we can use the normalize_affy_geno_cluster.pl program in the PennCNV-Affy package to calculate the LRR and BAF values:

```
perl PennCNV/affy/bin/normalize_affy_geno_cluster.pl
-locfile PennCNV/affy/lib/affygw6.hg38.pfb -out output/
```

Table 3
An example of the canonical genotype clustering file

probeset_id	r_aa	r_ab	r_bb	theta_aa	theta_ab	theta_bb	count_aa	count_ab	count_bb
SNP_A-2131660	3627.801	4198.245	3745.156	0.1512	0.5249	0.8893	3	57	78
SNP_A-1967418	988.5844	988.5844	892.0622	0.1729	0.4729	0.7819	0	38	101
SNP_A-1969580	3853.905	3853.905	3800.266	0.1985	0.4985	0.8085	0	2	141
SNP_A-4263484	2577.472	2527.32	2102.064	0.1021	0.4061	0.8183	11	74	49
SNP_A-1978185	1348.541	2116.979	2116.979	0.1746	0.6171	0.9171	142	7	0

```
gw6.LRR_BAF.txt output/gw6.genocluster output/quant-
norm.pm-only.med-polish.expr.summary.txt
```

The above command generates LRR and BAF values using the previously generated summary file (quant-norm.pm-only.med-polish.expr.summary.txt) and the clustering file (gw6.genocluster) in the output/ directory. A tab-delimited file named gw6.LRR_BAF.txt will be generated, which contains LRR and BAF values for each SNP and each sample. After this file is generated, we need to split this file into individual signal intensity files (one file for each sample).

We can use the kcolumn.pl program in PennCNV main package to split the gw6.LRR_BAF.txt file. An example is given below:

```
mkdir output/individual_signal_intensity_files/
perl PennCNV/kcolumn.pl output/gw6.LRR_BAF.txt split 2
--tab  --heading  3 --name_by_header  --output output/
individual_signal_intensity_files/gw6
```

The separated signal intensity files will be written in the output/individual_signal_intensity_files/ directory, each file with a prefix "gw6". These files can be used for CNV calling.

3.2.2 Preparation of Input List File

Although the signal file names can be provided in command line, the -list argument in PennCNV can take a list file that gives all file names to be processed. When calling CNV for each individual, the list file should contain one file name per line. When calling CNV for trios (using -trio argument or -joint argument), the list file should contain three file names per line separated by the tab character. When calling CNV for quartets, the list file contains four file names per line separated by tab character.

3.2.3 Preparation of PFB Files

The PFB (Population frequency of B allele) file supplies the PFB information for each marker, and gives the chromosome coordinates information to PennCNV. It is a tab-delimited text file with four columns, representing marker name, chromosome, position, and PFB values. When PFB value is 2, it means that the marker is a CN marker without polymorphism. An example of PFB file is shown in Table 4.

Table 4
An example of PFB file

Name	Chr	Position	PFB
rs300773	2	105035	0.816649899396378
rs2126131	2	119028	0.811015664477009
CN2000	2	120357	2

When reading the signal intensity file, PennCNV will only process markers annotated in the PFB file. Therefore, if we want to remove some markers from CNV analysis due to various reasons (such as being located within segmental duplication region, or within pseudo-autosomal region), we can simply remove these markers from the PFB file, without changing the signal intensity file per se. Similarly, if we want to call CNV on a different genome assembly (e.g., GRCh38 versus GRCh37), we can simply change the PFB file to reflect the new chromosome coordinates, without the need to change signal intensity files. Users can generate their own PFB file from a collection of signal intensity files (preferably more than 500 files). Assuming we already have enough signal intensity files, we can use the following command to generate the PFB file:

```
perl PennCNV/compile_pfb.pl -list input_list_file
-output output.pfb
```

In the command above, the input_list_file is the file that contains all the names of input signal intensity files, as described in Subheading 3.2.2. If the command runs correctly, it will generate an output file (output.pfb). The file should contain four columns as described above.

3.2.4 Preparation of GCModel File

The GCModel file specifies the GC content of the 1 Mb genomic region surrounding each marker (500 kb each side). It is used by the -gcmodel argument in PennCNV, and has been useful to salvage samples affected by genomic waves [33]. An example of GCModel file is shown in Table 5.

Note that the second and third columns are not used by PennCNV, since the information is already provided in the PFB file. The GC values range from 0 to 100, indicating the percentage of G or C base pairs in each region surrounding the marker. The GCModel file for several arrays is available on the PennCNV website. The PennCNV package also provides a script (cal_gc_snp.pl) to generate the GCModel file. To generate the GCModel file, we need the PFB file to provide the SNP information and a text-format gc5base file to provide the GC content of each 5kb window in the reference genome. The gc5base file can be downloaded from UCSC or be prepared as described in **Note 8**. The latest version of PennCNV provided gc5Base files for human reference genome

Table 5
An example of GCModel file

Name	Chr	Position	GC
rs6796976	3	111717110	40.026
rs1664136	3	135317005	40.173
rs11824188	11	20069295	42.580

hg18, hg19 and hg38. Assuming we have a PFB file named example.pfb and the chromosomal coordinates are based on human genome hg38, we can use the following command to download the gc5base file and generate the GCModel file:

```
gunzip PennCNV/gc_file/hg38.gc5Base.txt.gz
sort -k 2,2 -k 3,3n PennCNV/gc_file/hg38.gc5Base.txt >
PennCNV/gc_file/hg38.gc5Base.sort.txt
perl PennCNV/cal_gc_snp.pl PennCNV/gc_file/hg38.
gc5Base.sort.txt  PennCNV/example/example.pfb -output
PennCNV/example/example.gcmodel
```

If the command runs correctly, it will generate an output file (example.gcmodel) in the PennCNV/example/ directory. The file should contain four columns as described above.

3.3 Detection of CNVs

The PennCNV package provides some example data sets to test the program installation and to demonstrate the usage of PennCNV. In the PennCNV/example/ directory, we will see several files there. Among them, the father.txt, mother.txt and offspring.txt are three signal intensity files with signal values (to keep the file size small, only a few chromosomes are included in these files). In addition, there are also an HMM file (example.hmm) and a PFB file (example.pfb). In this section, we will use these example data sets to show how to detect CNVs using PennCNV.

3.3.1 CNV Detection on Individuals

Suppose we are already inside the PennCNV/example directory. We can use the following command to detect CNVs for the three individuals:

```
perl ../detect_cnv.pl -test -hmm example.hmm -pfb ex-
ample.pfb -conf -log example.rawcnv.log -out example.
rawcnv  father.txt mother.txt offspring.txt
```

In the command line, we have several arguments: the -test argument tells the program to generate CNV calls, while the -hmm and -pfb arguments specify the HMM model file and the PFB file; the -log argument specifies the file to store log information, and the -out argument specifies the file to store output CNV. father.txt, mother.txt, and offspring.txt are signal intensity files for each individual. The program will usually finish in a few minutes in a typical modern computer (*see* **Note 9**).

We can also store all the signal intensity files in one list file and then use the -list argument in the command line. This is especially useful when you have many input intensity files. As described above, when calling CNV for each individual, the list file should contain one file name per line. Assuming that the list file is named inputlist, we can use the following command to perform CNV calling:

```
perl ../detect_cnv.pl -test -hmm example.hmm -pfb ex-
ample.pfb -conf -log example.rawcnv.log -out example.
rawcnv  -list inputlist
```

The output CNV calls are stored in the example.rawcnv file. The first a few lines are shown in Table 6. The first column is the chromosome region, which is based on the chromosomal coordinates stored in the PFB file. The second column and the third column describe how many SNPs are contained within the CNV and the length of the CNV. The fourth column is the HMM state and the actual copy number (CN) of the CNV call. The CN refers to the actual integer copy number estimates, and the diploid copy number is 2. So for autosome, CN = 0 or 1 means there is a deletion and CN ≥ 3 means there is a duplication. For chrX or chrY in males, CN = 1 is the normal copy number and CN = 0 means a deletion. The fifth, sixth and seventh columns specify the input signal intensity file name, the starting marker identifier, and the ending marker identifier in the CNV, respectively (*see* **Note 10**).

Since the GC content around the SNP may have some effect on the signal intensity and create "genomic waves" [33], it is sometimes necessary to adjust the GC-wave to reduce the false positive calls. PennCNV provided a wave adjustment procedure via the -gcmodel argument. This procedure requires a GCModel file, the preparation of which is described in Subheading 3.2.4. Assuming we already prepared the GCModel file (named example.gcmodel), we can use the following command to do CNV calling with GC adjustment:

```
perl ../detect_cnv.pl -test -hmm example.hmm -pfb ex-
ample.pfb -log example.adjusted.log -out example.ad-
justed.rawcnv -gcmodel example.gcmodel -list inputlist
```

This will apply the GC-model specified in example.gcmodel for signal adjustment, before generating CNV calls.

3.3.2 Trio-Based CNV Calling

The family structure can be used for generating more accurate CNV calls, since we can borrow and correlate CNV information from related family members that may share the same CNV region. Suppose we already generated the CNV calls for the three family members (father, mother, and offspring) as described in the previous section. We can use the following command to perform a trio-based CNV calling:

```
perl ../detect_cnv.pl -trio -hmm example.hmm -pfb ex-
ample.pfb -cnv example.rawcnv -log example.triocnv.log
-out example.triocnv father.txt mother.txt offspring.
txt
```

In the above command, the -trio argument specifies that we want to use family-based CNV detection algorithm to update CNV

Table 6
An example of CNV call file

chr3:37940970-37944758	numsnp=3	length=3789	state2, cn=1	father.txt	startsnp=rs9837352	endsnp=rs9844203	conf=15.133
chr3:75379524-75519068	numsnp=7	length=139,545	state2, cn=1	father.txt	startsnp=rs4677005	endsnp=rs2004089	conf=26.863
chr11:81792950-81806219	numsnp=9	length=13,270	state2, cn=1	father.txt	startsnp=rs7947005	endsnp=rs12293984	conf=35.081
chr20:10511631-10583260	numsnp=10	length=71,630	state2, cn=1	father.txt	startsnp=rs8114269	endsnp=rs682562	conf=39.413

status for a father–mother–offspring trio. The -cnv argument specifies the prior CNV calls generated in individual-based calling step. The three .txt files in command line represent signal intensity file for father, mother and offspring, respectively. The output will be written to example.triocnv (*see* **Note 11**).

The first a few lines of the output CNV call file is shown in Table 7. As we can see, the trio-based CNV file contains two extra fields: the eighth field indicates that the input file offspring.txt is offspring in the trio family, while the ninth field tells us that the HMM states for the trio are 3 (normal), 3 (normal), and 2 (one-copy deletion) at this genomic region, respectively (*see* **Note 12**).

3.3.3 Joint CNV Calling

Unlike the trio-based calling algorithm, which uses posterior validation on individual-based CNV calls, the joint-calling algorithm in PennCNV generates CNV calls in one single step for three individuals in a family [34]. The joint CNV calling algorithm has better performance than the current family-based CNV calls, especially in resolving the correct CNV boundaries and for reducing false negative rates on very small CNV calls. However, it is substantially slower than trio-calling algorithm, and may take several hours for a single trio in a typical modern computer. To use this algorithm, we can specify -joint argument, rather than -trio argument in the command line. For example:

```
perl ../detect_cnv.pl -joint -hmm example.hmm -pfb ex-
ample.pfb -log example.joint.rawcnv.log -out example.
joint.rawcnv father.txt mother.txt offspring.txt
```

As we can see from the command line above, unlike the trio-based algorithm, the joint CNV calling algorithm does not require an existing CNV file generated by individual-based calling algorithm as input (*see* **Note 13**).

3.3.4 Merging Adjacent CNV Calls

Sometimes PennCNV may generate several small close-spaced CNV calls for a large CNV. Therefore, we need to examine the CNV calls and merge adjacent calls together if they are close to each other and share the same copy number. We can use the clean_cnv.pl program to merge the adjacent CNV calls. By default, it will merge two nearby CNV calls if the gap between them is less than 20% of the total length of the two calls plus the gap region. For example, we can use the following command to merge the calls in the example.rawcnv file:

```
perl ../clean_cnv.pl combineseg  -fraction 0.2 -bp
-signalfile example.pfb  example.rawcnv  > example.
rawcnv.merge
```

In the above command, the combineseg argument specifies that the task is to combine nearby segments (i.e., merge calls). The

Table 7
An example of trio-based CNV call file

chr3:3957986-4054960	numsnp=50	length=96,975	state2, cn=1	offspring.txt	startsnp=rs11716390	endsnp=rs17039742	offspring	triostate=332
chr3:37940970-37944758	numsnp=3	length=3789	state2, cn=1	father.txt	startsnp=rs9837352	endsnp=rs9844203	father	triostate=233
chr3:75379524-75519068	numsnp=7	length=139,545	state2, cn=1	father.txt	startsnp=rs4677005	endsnp=rs2004089	father	triostate=233
chr11:549119-558884	numsnp=4	length=9766	state5, cn=3	mother.txt	startsnp=rs4963136	endsnp=rs2061586	mother	triostate=355

-fraction 0.2 argument specifies that the fraction threshold is 0.2, and the -bp argument specifies that the fraction is measured by base pair length, rather than the number of SNP markers.

3.4 Filtering CNV Calls

The raw CNV calls often need to be filtered to keep a specific subset of calls for further analysis. In the PennCNV package, the filter_cnv.pl program can filter CNV calls based on various criteria, including both call-level and sample-level criteria.

3.4.1 Filtering CNV Calls Based on Call-Level Criteria

If we only want to retain CNV calls that are larger than 50 kb and contain more than 10 SNPs, we can use the following command to filter the calls:

```
perl ../filter_cnv.pl -numsnp 10 -length 50k example.
rawcnv -out example.snp10.length10k.cnv
```

If the command runs correctly, the CNV calls meeting the specified criteria will be written to the file example.snp10.length10k.cnv. Note that the -numsnp argument works for both SNP markers and CN markers without polymorphism.

3.4.2 Filtering CNV Calls Based on Sample-Level Criteria

We can use the filter_cnv.pl program to identify low-quality samples from a genotyping experiment, and eliminate them from future analysis. This analysis requires the LOG file used in CNV calling. Low-quality samples often have large LRR_SD (standard deviation of LRR values in autosomes) values. Therefore, we can filter the low-quality samples using this criterion. For example, if we want to remove samples of which the LRR_SD value is larger than 0.3, we can use the following command:

```
perl ../filter_cnv.pl example.rawcnv -qclogfile example.
rawcnv.log -qclrrsd 0.3 -qcpassout example.qcpass -qc-
sumout example.qcsum -out example.goodcnv
```

This command will analyze the log file (example.rawcnv.log), find all samples with LRR_SD less than 0.3, then write these samples to the example.qcpass file, write the CNV calls of these samples to the example.goodcnv file, and write the QC summary for all samples to the example.qcsum file. Generally, users can examine the relationship between LRR_SD and number of calls in a given cohort, and select a threshold manually that reach a good compromise between including as much samples as possible while reducing false positive calls; a value between 0.25 and 0.3 is used in many studies.

We also recommend to use the -qcnumcnv argument in the command to filter out samples that have too many CNV calls. For example, -qcnumcnv 100 would treat any samples with > 100 CNV calls as low quality samples and eliminate them from analysis.

The .qcsum file contains several QC summary statistics for all samples. An example of the file is shown in Table 8. LRR_mean

Table 8
An example of qcsum file

File	LRR_mean	LRR_median	LRR_SD	BAF_mean	BAF_median	BAF_SD	BAF_drift	WF	NumCNV
mother.txt	0.0039	0	0.1374	0.5044	0.5	0.0418	0.00014	0.01	2
father.txt	0.0027	0	0.1335	0.5063	0.5	0.039	0.000037	0.0184	4
offspring.txt	0.0028	0	0.1263	0.5045	0.5	0.0429	0.000293	−0.0171	4

and LRR_median represent the mean and median of LRR values of the sample. BAF_mean, BAF_median, and BAF_SD are the mean, median and standard deviation of BAF values of the sample. BAF_drift is the fraction of markers with BAF values between 0.2 and 0.25 or between 0.75 and 0.8 for autosomes; it is a measure of random noises in the data and can be useful to detect sample mix-up or the use of non-optimal clustering files in LRR/BAF signal generation. WF is the wave factor, which measures the magnitude and directionality of genomic waves of LRR [33]. These statistics are calculated based on autosomes. The .qcsum file is a tab-delimited file and can be easily loaded into Excel for plots and histograms. For example, it is often informative to plot the number of CNV calls and the LRR_SD measure to find a good threshold to use for filtering for a particular data set. Figure 2a is a scatter plot showing the number of CNV calls versus LRR_SD of a cohort. As we can see, most of the samples have an LRR_SD value less than 0.4 and have number of CNV calls less than 100. However, there are some outliers that have very large LRR_SD values or very large numbers of CNV calls, which should probably be flagged or even excluded from downstream analysis (visualization of signal intensity values at specific CNV regions would also be important as well). We also plotted the histogram of LRR_SD values and number of CNV calls of the cohort in Fig. 2b, c, respectively. In this case, we can arbitrarily set 0.4 as the threshold of LRR_SD and 100 as the threshold of number of CNV calls. These thresholds will differ by array platforms and genotyping batches.

Recently, Macé A et al. shared an article on quality control of CNV calls detected by PennCNV [35]. They defined a new score (QS) to estimate the probability of a CNV called by PennCNV to being a consensus call (i.e., can be detected by other CNV callers). QS combines multiple sample parameters provided by the PennCNV. They wrapped up the QS calculation in a pipeline designed to run CNV trait associations. The pipeline is available online at http://goo.gl/T6yuFM, and it may be a good source for PennCNV users to perform quality control of CNV calls.

3.4.3 Removing Spurious CNV Calls in Specific Genomic Regions

Several genomic regions such as immunoglobulin regions and centromeric/telomeric regions are known to harbor spurious CNV calls, which should be eliminated before analysis. We can use the scan_region.pl program in the PennCNV package to remove CNV calls in specific genomic regions. For example, we can use the following command to remove CNV calls that overlap with immunoglobulin regions:

```
perl ../scan_region.pl example.rawcnv imm_region -min-
queryfrac 0.5 > example.rawcnv.imm_region
```

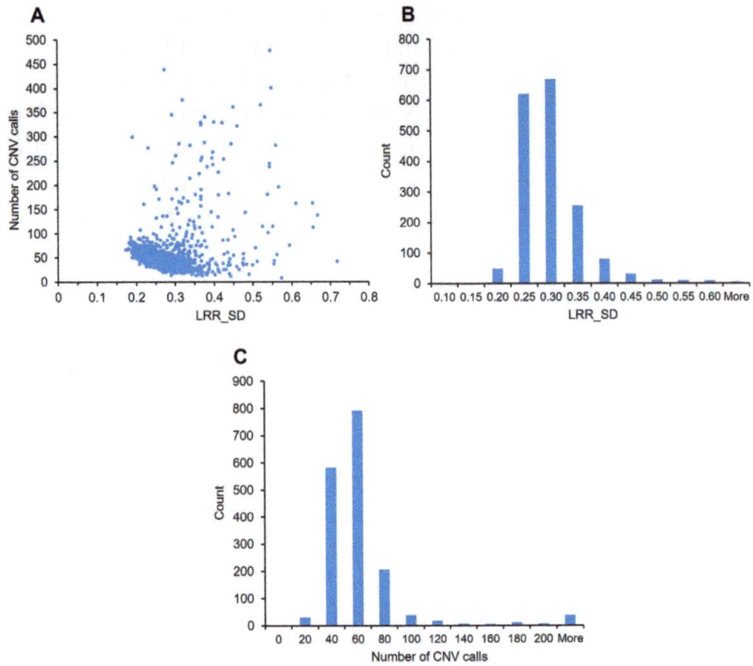

Fig. 2 The relationship between LRR_SD and the number of CNV calls for all samples in a cohort. (**a**) Scatter plot showing LRR_SD versus the number of CNV calls. (**b**) Histogram of LRR_SD values. (**c**) Histogram of the number of CNV calls

```
grep -v -f  example.rawcnv.imm_region example.rawcnv >
example.rawcnv.clean
```

This command scans the CNV call file (example.rawcnv) against known immunoglobulin regions (stored in file imm_region), and output CNV calls that overlap with the immunoglobulin regions to the example.rawcnv.imm_region file. The "-minqueryfrac 0.5" argument specifies that at least 50% of the length in the CNV call must overlap with the immunoglobulin region. Then the grep program is used to remove the calls in the example.rawcnv.imm_region file from the original call file and generate a cleaned call file (example.rawcnv.clean). The imm_region file contains immunoglobulin regions, in the format of "chr1:1000-2000", and each line contains one region.

3.5 Annotation of CNV Calls

3.5.1 Finding Overlapping/Neighboring Genes of CNV Calls

One of the most common tasks for CNV annotation is to identify overlapping or neighboring genes. We need to download the refGene annotation files (refGene.txt.gz) and then use the scan_region.pl program to find the overlapping calls. Assuming reference genome is hg38, we can run the following commands:

```
wget http://hgdownload.cse.ucsc.edu/goldenPath/hg38/
database/refGene.txt.gz
gunzip -c refGene.txt.gz > hg38.refGene.txt
```

```
perl ../scan_region.pl example.rawcnv hg38.refGene.txt
-refgene_flag -name2_flag > example.cnv.overlap_hg38
```

The output file contains two additional columns to each line of the example.rawcnv file. The first column represents the gene symbols and the second column indicates the distance between CNV and gene. If the CNV overlaps with a gene, the distance would be zero. If the CNV does not overlap with any gene, a "NOT_FOUND" notation will be shown for the corresponding CNVs. The first few lines of the output file (example.cnv.overlap_hg38) are shown in Table 9.

If we want to find neighboring genes, we can use the -expandmax argument:

```
perl ../scan_region.pl example.rawcnv hg38.refGene.txt
-refgene_flag -name2_flag  -expandmax 5m > example.cnv.
overlap_hg38.expand_5m
```

This will expand the CNV up to five megabases in both directions and then try to find neighboring genes. Only the closest gene to the CNV will be written to output, while this closest gene might be located to the left or right side of the CNV. To find only left genes, we can use the -expandleft 5m argument.

3.5.2 Finding Overlapping/Neighboring Exons of CNV Calls

CNVs that overlap exons may severely affect gene functions. We can run the scan_region.pl program and specify the -refexon argument, instead of the -refgene argument to find exonic overlaps:

```
perl ../scan_region.pl example.rawcnv hg38.refGene.
txt -refgene_flag -name2_flag -refexon > example.cnv.
hg38_exon
```

The CNV calls without exonic overlap will have "NOT_FOUND" appended to the end of the line. Therefore, we can use the following command to remove the nonexonic CNV calls:

```
grep -v NOT_FOUND example.cnv.hg38_exon > example.cnv.
hg38_exon_found
```

3.6 Visualization of CNV Calls

It is often helpful to visually examine CNV calls to judge whether they are reliable or not. PennCNV provides a convenient way to generate image files for CNV calls automatically. For example, if we want to plot the CNV calls of the offspring, we can run the following command:

```
perl ../visualize_cnv.pl -format plot -signal off-
spring.txt example.rawcnv
```

This command will read both the CNV call file (example.rawcnv) and the signal intensity file (offspring.txt), and then plot the signal intensities (LRR/BAF) for all the CNV calls that are detected

Table 9
An example of CNV annotation file

chr3:37940970-37944758	numsnp=3	length=3789	state2, cn=1	father. txt	startsnp=rs9837352	endsnp=rs9844203	conf=15.133	CTDSPL
chr3:75379524-75519068	numsnp=7	length=139,545	state2, cn=1	father. txt	startsnp=rs4677005	endsnp=rs2004089	conf=26.863	FAM86DP,LINC02018
chr11:81792950-81806219	numsnp=9	length=13,270	state2, cn=1	father. txt	startsnp=rs7947005	endsnp=rs12293984	conf=35.081	NOT_FOUND
chr20:10511631-10583260	numsnp=10	length=71,630	state2, cn=1	father. txt	startsnp=rs8114269	endsnp=rs682562	conf=39.413	SLX4IP

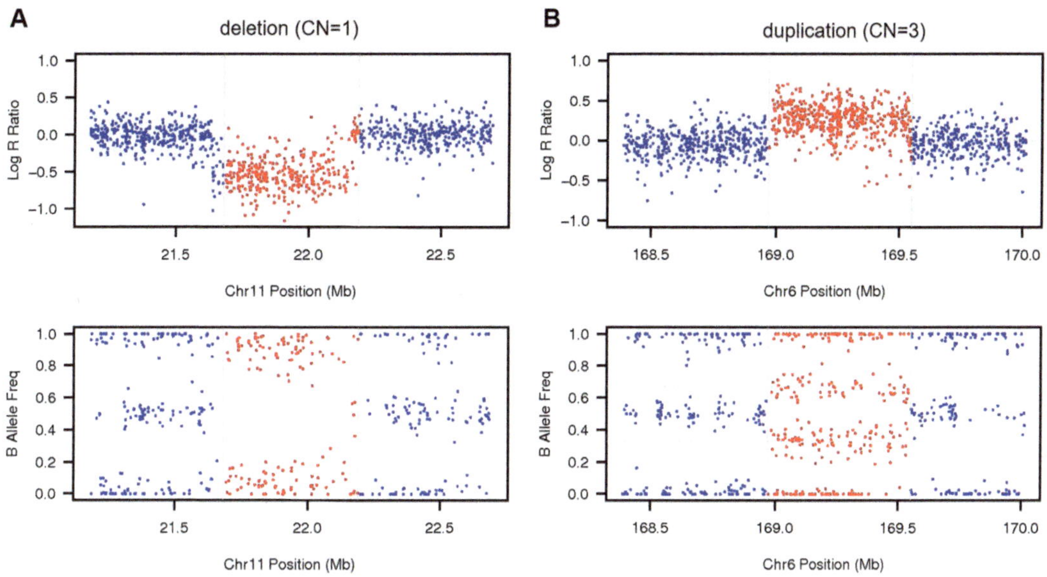

Fig. 3 Plot of LRR and BAF values of two CNV calls. (**a**) LRR and BAF values of a deletion (CN = 1) are shown in upper and lower panels, respectively. (**b**) LRR and BAF values of a duplication (CN = 3) are shown in upper and lower panels, respectively. The red dots represent the markers inside the CNV calls

from the specified signal intensity file (offspring.txt). The plotting function requires R to be installed. The output are image files in JPG formats or PDF formats.

We plotted one deletion example (CN = 1) and one duplication example (CN = 3) in Fig. 3. By default, the CNV region, as well as the left side and right side region with identical sizes, is included in the figure. The CNV region is marked by two gray vertical lines. In normal regions (Fig. 3a blue regions), the log R ratios are around zero and B allele frequencies are around three values: 0.0, 0.5, and 1.0. In the deletion region with one copy (Fig. 3a, red dots), the log R ratios drop to about −0.5 and the BAF values are around two values: 0.0 and 1.0. This is because in the one-copy region, there is only one allele and the genotype can only be A or B. In the duplication region with three copies (Fig. 3b, red dots), the Log R Ratios increase to about +0.5 and the BAF values scatter around four values: 0.0, 0.33, 0.67, and 1.0. This is because in the duplication region, there are three alleles and the genotype can only be AAA, AAB, ABB, or BBB.

4 Notes

1. You can add the PennCNV directory into the PATH environmental variable in your operating system, so that all PennCNV scripts can be executed directly by typing the name of the command.

2. If you have problems installing PennCNV in your operating system, it is perhaps due to the incompatibilities of PennCNV's khmm module with certain Perl installations in the operating system. To solve this issue, you can use perlbrew to install a different version of Perl (such as 5.14.2); for example, you can use the command "perlbrew install perl-5.14.2 --as perl-5.14.2-PIC -Accflags=-fPIC" to install Perl 5.14.

   If you are using Windows, we recommend that you first download and install 32-bit Perl 5.8.8 and then use PennCNV directly. In this case, there is no need for compilation because the .dll files for Perl 5.8.8 are already compiled and provided in the PennCNV package.

3. The Penn-Affy workflow can be adapted to other SNP array platforms. For example, Joseph T. et al. applied the Penn-Affy workflow on the Perlegen 600K platform [36]. The generate_affy_geno_cluster.pl program in the Penn-Affy package requires three input files: a genotype call file, a confidence file that contains the confidence values of the genotype calls, a signal intensity file that contains normalized signal intensities of A and B alleles and a location file that contains genomic locations of markers (e.g., a PFB file, described in Subheading 3.2.3). For Affymetrix arrays, the first three files can be generated by Affymetrix Power Tools. Users of other platforms can generate the required data values using their platform-specific tools and then reformat the data into the file formats as described above. The signal intensity values can be transformed into log-scale. After generating the four input files, users can generate the canonical cluster file using generate_affy_geno_cluster.pl and then generate the LRR and BAF values using normalize_affy_geno_cluster.pl (*see* Subheading 3.2.1, **steps 4** and **5**).

4. We can use the following commands to download and unzip the example data set:

```
mkdir raw_data
cd raw_data
wget ftp://ftp.ncbi.nlm.nih.gov/geo/series/
GSE15nnn/GSE15826/suppl/GSE15826_RAW.tar
tar xf GSE15826_RAW.tar
gunzip *.gz
```

5. For a typical modern computer, the command should take less than 1 day for 1000–2000 CEL files. It is very important to check that the APT programs finish completely, before proceeding to next steps. Check the LOG files to see whether it reports a success.

6. We need to use at least 500 CEL files to generate a high-quality clustering file. If only a few CEL files are available, users can skip this step and use the default canonical clustering file in the PennCNV-Affy package for the identical array (if available),

but in this case the CNV calls may be less reliable. Examples of such clustering files are: hapmap.genocluster for Genome-Wide SNP Array 6.0, agre.genocluster for Genome-Wide SNP Array 5.0, and affy500k.nsp.genocluster/affy500k.sty.genocluster for Mapping 500K Array Set.

7. If the sex information for some CEL file is not known, you do not need to include them in the cel_sex_file. The birdseed.report.txt file that was generated in the previous contains a field named computed_gender. Therefore, we can use the following command to generate the cel_sex_file:

```
cut -f 1-2 birdseed.report.txt  | grep male > cel_
sex_file
```

8. For some reference genomes, the text-format gc5Base file is not officially provided by UCSC. In this case, we can prepare the gc5Base file by the following steps.

Step 1, download two tools provided by UCSC:

```
wget http://hgdownload.cse.ucsc.edu/admin/exe/
linux.x86_64/faToTwoBit
chmod +x faToTwoBit
wget http://hgdownload.cse.ucsc.edu/admin/exe/
linux.x86_64/hgGcPercent
chmod +x hgGcPercent
```

faToTwoBit and hgGcPercent are binary files precompiled by UCSC and are free for academic, nonprofit, and personal use. A license may be required for commercial use.

Step 2, convert the reference FASTA file to .2bit file (assuming the reference file is hg38.fa):

```
./faToTwoBit hg38.fa  hg38.2bit
```

Step 3, generate GC content file in Wiggle format:

```
./hgGcPercent -wigOut -doGaps -file=stdout -win=5120
hg38 hg38.2bit > hg38.gc.wig
```

Step 4, generate gc5Base.txt file using the script provided in PennCNV/gc_file directory:

```
PennCNV/gc_file/wig2gc5base hg38.gc.wig > hg38.
gc5Base.txt
```

9. By default, only autosome CNVs will be detected, the –chrx argument can be used to generate CNV calls on (and only on) chromosome X. The CNV calling for chrX is slightly different from that of autosomes. It is highly recommended to use the -sexfile argument to supply gender annotation for all genotyped samples. The sexfile is a two-column file, with the first

Table 10
An example of sexfile

father.txt	male
mother.txt	female
offspring.txt	male

column being signal file names, and the second column being either "male" or "female". Table 10 shows an example of sexfile.

```
perl ../detect_cnv.pl -test -hmm example.hmm -pfb ex-
ample.pfb -log example.rawcnv.log -out example.rawcnv
-list inputlist -chrx -sexfile sexfile.txt
```

If sex for a sample is not provided in sexfile, or if -sexfile is not specified, PennCNV will try to predict the gender of the sample, based on BAF heterozygosity rate of chrX markers, but such predictions may not be reliable for some arrays. Next, PennCNV will adjust the LRR values such that females have median LRR at 0 and males have median LRR at the same value as that for CN = 1 in the HMM file. After this step, the CNV calling is then applied in a similar way as autosomes.

10. As of June 2008, the -medianadjust argument is turned on by default in the program to reduce false positive duplication calls for problematic samples. The effect is that the BAF_median measure for all samples is automatically adjusted to be 0.5. Users can turn off the argument by specifying -nomedianadjust. This is important when calling CNVs on a signal intensity file that contains data only on a specific genomic region rather than a whole genome.
11. If we have multiple trio families, we can generate a listfile, which contains three file names per line (i.e., one family per line), to process multiple trios simultaneously. It is important that the signal intensity file names in the command line (or in the listfile) are identical to the file names listed in the fifth column of the CNV call file (e.g., example.rawcnv) so that PennCNV can recognize the correct signal intensity file of each call.
12. If the family has two children, then the -quartet argument can be used for CNV calling. Accordingly, four file names should be supplied in the command line, or given in each line of the list file, representing father, mother, child 1, and child 2, respectively. PennCNV cannot generate calls on a pair of parents and three or more children; instead, the user need to split the family into trios and quartets for CNV calling, and then combine the CNV calls together into consensus calls.

13. The joint CNV calling algorithm only supports trio families. For complex nuclear families, it is better to use the -trio and -quartet operations described in Subheading 3.3.2.

References

1. Feuk L, Carson AR, Scherer SW (2006) Structural variation in the human genome. Nat Rev Genet 7(2):85–97. https://doi.org/10.1038/nrg1767
2. Zarrei M, MacDonald JR, Merico D et al (2015) A copy number variation map of the human genome. Nat Rev Genet 16(3):172–183. https://doi.org/10.1038/nrg3871
3. Sudmant PH, Rausch T, Gardner EJ et al (2015) An integrated map of structural variation in 2,504 human genomes. Nature 526(7571):75–81. https://doi.org/10.1038/nature15394
4. Mills RE, Walter K, Stewart C et al (2011) Mapping copy number variation by population-scale genome sequencing. Nature 470(7332):59–65. https://doi.org/10.1038/nature09708
5. Zhang F, Gu W, Hurles ME et al (2009) Copy number variation in human health, disease, and evolution. Annu Rev Genomics Hum Genet 10:451–481. https://doi.org/10.1146/annurev.genom.9.081307.164217
6. Girirajan S, Campbell CD, Eichler EE (2011) Human copy number variation and complex genetic disease. Annu Rev Genet 45:203–226. https://doi.org/10.1146/annurev-genet-102209-163544
7. Weischenfeldt J, Symmons O, Spitz F et al (2013) Phenotypic impact of genomic structural variation: insights from and for human disease. Nat Rev Genet 14(2):125–138. https://doi.org/10.1038/nrg3373
8. Watson CT, Marques-Bonet T, Sharp AJ et al (2014) The genetics of microdeletion and microduplication syndromes: an update. Annu Rev Genomics Hum Genet 15:215–244. https://doi.org/10.1146/annurev-genom-091212-153408
9. Zack TI, Schumacher SE, Carter SL et al (2013) Pan-cancer patterns of somatic copy number alteration. Nat Genet 45(10):1134–1140. https://doi.org/10.1038/ng.2760
10. Beroukhim R, Mermel CH, Porter D et al (2010) The landscape of somatic copy-number alteration across human cancers. Nature 463(7283):899–905. https://doi.org/10.1038/nature08822
11. Carter NP (2007) Methods and strategies for analyzing copy number variation using DNA microarrays. Nat Genet 39(7 Suppl):S16–S21. https://doi.org/10.1038/ng2028
12. Pinto D, Darvishi K, Shi X et al (2011) Comprehensive assessment of array-based platforms and calling algorithms for detection of copy number variants. Nat Biotechnol 29(6):512–520. https://doi.org/10.1038/nbt.1852
13. Venkatraman ES, Olshen AB (2007) A faster circular binary segmentation algorithm for the analysis of array CGH data. Bioinformatics 23(6):657–663. https://doi.org/10.1093/bioinformatics/btl646
14. Olshen AB, Venkatraman ES, Lucito R et al (2004) Circular binary segmentation for the analysis of array-based DNA copy number data. Biostatistics 5(4):557–572. https://doi.org/10.1093/biostatistics/kxh008
15. Price TS, Regan R, Mott R et al (2005) SW-ARRAY: a dynamic programming solution for the identification of copy-number changes in genomic DNA using array comparative genome hybridization data. Nucleic Acids Res 33(11):3455–3464. https://doi.org/10.1093/nar/gki643
16. Cooper GM, Zerr T, Kidd JM et al (2008) Systematic assessment of copy number variant detection via genome-wide SNP genotyping. Nat Genet 40(10):1199–1203. https://doi.org/10.1038/ng.236
17. Peiffer DA, Le JM, Steemers FJ et al (2006) High-resolution genomic profiling of chromosomal aberrations using Infinium whole-genome genotyping. Genome Res 16(9):1136–1148. https://doi.org/10.1101/gr.5402306
18. Wang K, Li M, Hadley D et al (2007) PennCNV: an integrated hidden Markov model designed for high-resolution copy number variation detection in whole-genome SNP genotyping data. Genome Res 17(11):1665–1674. https://doi.org/10.1101/gr.6861907
19. Colella S, Yau C, Taylor JM et al (2007) QuantiSNP: an Objective Bayes Hidden-Markov Model to detect and accurately map copy number variation using SNP genotyping data. Nucleic Acids Res 35(6):2013–2025. https://doi.org/10.1093/nar/gkm076
20. Zhang X, Du R, Li S et al (2014) Evaluation of copy number variation detection for a SNP array platform. BMC Bioinformatics 15:50. https://doi.org/10.1186/1471-2105-15-50
21. Marenne G, Rodriguez-Santiago B, Closas MG et al (2011) Assessment of copy number

variation using the Illumina Infinium 1M SNP-array: a comparison of methodological approaches in the Spanish Bladder Cancer/EPICURO study. Hum Mutat 32(2):240–248. https://doi.org/10.1002/humu.21398

22. Dellinger AE, Saw SM, Goh LK et al (2010) Comparative analyses of seven algorithms for copy number variant identification from single nucleotide polymorphism arrays. Nucleic Acids Res 38(9):e105. https://doi.org/10.1093/nar/gkq040
23. Sanders SJ, He X, Willsey AJ et al (2015) Insights into autism spectrum disorder genomic architecture and biology from 71 risk loci. Neuron 87(6):1215–1233. https://doi.org/10.1016/j.neuron.2015.09.016
24. Huang AY, Yu D, Davis LK et al (2017) Rare copy number variants in NRXN1 and CNTN6 increase risk for tourette syndrome. Neuron 94(6):1101–1111 e1107. https://doi.org/10.1016/j.neuron.2017.06.010
25. Marshall CR, Howrigan DP, Merico D et al (2017) Contribution of copy number variants to schizophrenia from a genome-wide study of 41,321 subjects. Nat Genet 49(1):27–35. https://doi.org/10.1038/ng.3725
26. Elia J, Glessner JT, Wang K et al (2011) Genome-wide copy number variation study associates metabotropic glutamate receptor gene networks with attention deficit hyperactivity disorder. Nat Genet 44(1):78–84. https://doi.org/10.1038/ng.1013
27. Green EK, Rees E, Walters JT et al (2016) Copy number variation in bipolar disorder. Mol Psychiatry 21(1):89–93. https://doi.org/10.1038/mp.2014.174
28. Rucker JJ, Tansey KE, Rivera M et al (2016) Phenotypic association analyses with copy number variation in recurrent depressive disorder. Biol Psychiatry 79(4):329–336. https://doi.org/10.1016/j.biopsych.2015.02.025
29. Glessner JT, Li J, Hakonarson H (2013) ParseCNV integrative copy number variation association software with quality tracking. Nucleic Acids Res 41(5):e64. https://doi.org/10.1093/nar/gks1346
30. McCarroll SA, Kuruvilla FG, Korn JM et al (2008) Integrated detection and population-genetic analysis of SNPs and copy number variation. Nat Genet 40(10):1166–1174. https://doi.org/10.1038/ng.238
31. Korn JM, Kuruvilla FG, McCarroll SA et al (2008) Integrated genotype calling and association analysis of SNPs, common copy number polymorphisms and rare CNVs. Nat Genet 40(10):1253–1260. https://doi.org/10.1038/ng.237
32. Staaf J, Vallon-Christersson J, Lindgren D et al (2008) Normalization of Illumina Infinium whole-genome SNP data improves copy number estimates and allelic intensity ratios. BMC Bioinformatics 9:409. https://doi.org/10.1186/1471-2105-9-409
33. Diskin SJ, Li M, Hou C et al (2008) Adjustment of genomic waves in signal intensities from whole-genome SNP genotyping platforms. Nucleic Acids Res 36(19):e126. https://doi.org/10.1093/nar/gkn556
34. Wang K, Chen Z, Tadesse MG et al (2008) Modeling genetic inheritance of copy number variations. Nucleic Acids Res 36(21):e138. https://doi.org/10.1093/nar/gkn641
35. Mace A, Tuke MA, Beckmann JS et al (2016) New quality measure for SNP array based CNV detection. Bioinformatics 32(21):3298–3305. https://doi.org/10.1093/bioinformatics/btw477
36. Glessner JT, Wang K, Sleiman PM et al (2010) Duplication of the SLIT3 locus on 5q35.1 predisposes to major depressive disorder. PLoS One 5(12):e15463. https://doi.org/10.1371/journal.pone.0015463

Chapter 2

Using SAAS-CNV to Detect and Characterize Somatic Copy Number Alterations in Cancer Genomes from Next Generation Sequencing and SNP Array Data

Zhongyang Zhang and Ke Hao

Abstract

Somatic copy number alterations (SCNAs) are profound in cancer genomes at different stages: oncogenesis, progression, and metastasis. Accurate detection and characterization of SCNA landscape at genome-wide scale are of great importance. Next-generation sequencing and SNP array are current technology of choice for SCNA analysis. They are able to quantify SCNA with high resolution and meanwhile raise great challenges in data analysis. To this end, we have developed an R package *saasCNV* for SCNA analysis using (1) whole-genome sequencing (WGS), (2) whole-exome sequencing (WES) or (3) whole-genome SNP array data. In this chapter, we provide the features of the package and step-by-step instructions in detail.

Key words Copy number variation, Somatic copy number alteration, SAAS-CNV, Next-generation sequencing, Whole-genome sequencing, Whole-exome sequencing, SNP array, Segmentation, Cancer genome

1 Introduction

Somatic copy number alterations (SCNAs) are a major type of DNA structural variations and ubiquitously exist in various cancer genomes at different stages. They play important roles in oncogenesis, progression and metastasis [1–4]. Accurate detection and characterization of SCNAs are essential in both research and clinical settings, especially for personalized cancer therapy [5]. The identification of SCNA is more complicated than copy number variant (CNV) detection in normal genome due to aneuploidy and heterogeneity of cancer cells [6]. Array-CGH [7] and SNP array [8] are commonly used high-throughput techniques to characterize SCNA profile at genome-wide scale. With the emergence of next-generation sequencing (NGS), such as whole-genome sequencing (WGS) and whole-exome sequencing (WES), over the last decade, the genome-wide SCNA profile can be quantified at unprecedented resolution and, simultaneously, at continuously

Derek M. Bickhart (ed.), *Copy Number Variants: Methods and Protocols*, Methods in Molecular Biology, vol. 1833, https://doi.org/10.1007/978-1-4939-8666-8_2,

decreasing cost [9, 10]. Meanwhile, these techniques generate a huge amount of data, demanding more powerful and sophisticated methods for data analysis. To this end, we have developed an SCNA analysis method SAAS-CNV [11] and implemented in an R package *saasCNV*. The SAAS-CNV has been compared to several state-of-the-art SCNA analysis methods based on a large breast cancer dataset from The Cancer Genome Atlas (TCGA) and achieved the best performance [12].

saasCNV package is applicable to WGS/WES data and whole-genome SNP array data generated from tumor and matched normal tissue. The main workflow of the pipeline is summarized in Fig. 1 [11]. From the sequencing (SNP array) platform, two types of information relevant to SCNA status are extracted at loci with heterozygous genotype: (1) total read depth (total fluorescent intensity) reflecting total copy number change; (2) allele specific read depth (allele specific intensity) reflecting allelic imbalance as a result of differential copy number changes upon the two alleles.

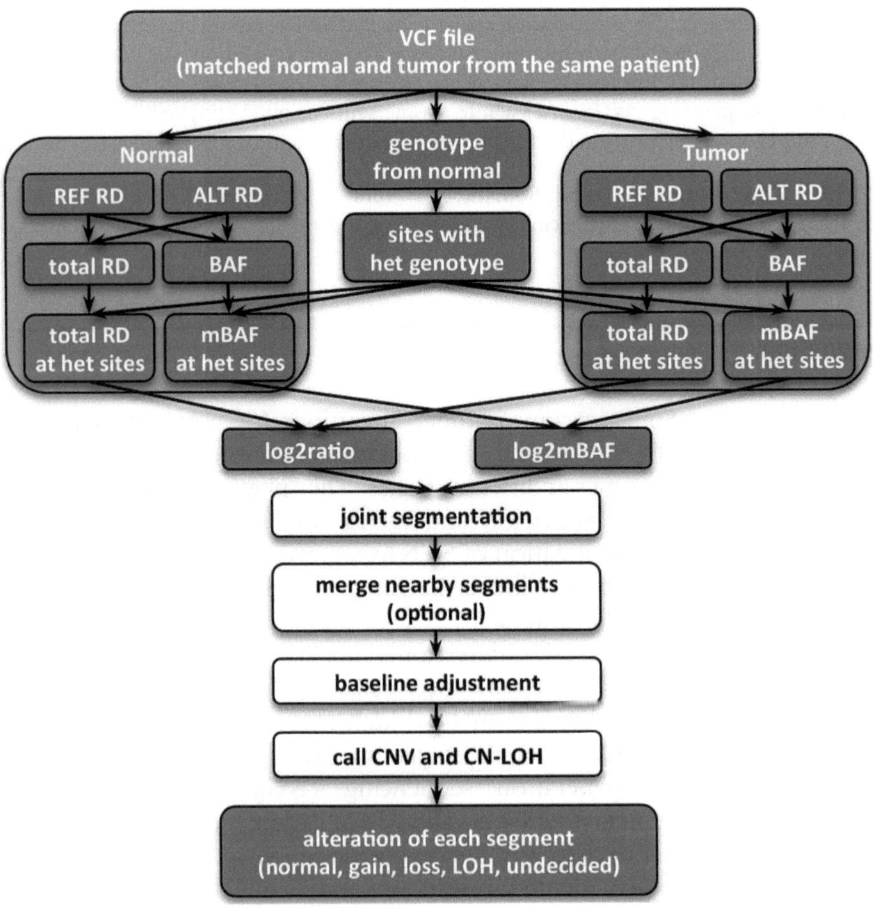

Fig. 1 The analysis workflow of *saasCNV* package. Picture was reproduced from [11] under the Creative Commons Attribution (CC BY) license

The two types of information also provide valuable clues for the inference of tumor ploidy and purity. These two types of signals are transformed into quantitative measures, called log2ratio and log2mBAF, respectively [11]. A joint analysis is then performed on these two signal dimensions in both segmentation and SCNA calling steps. In the segmentation step, the two signal sequences are jointly partitioned into segments of different sizes, each corresponding to a particular copy number status. In the following SCNA calling step, the algorithm identifies the baseline values for both log2ratio and log2mBAF dimensions, based on which the SCNA status is decided for each segment. The possible SCNA statuses include normal, loss, gain, copy-neutral loss of heterozygosity (CN-LOH) and "undecided" if no confident status can be decided.

For each tumor–normal pair, the *saasCNV* package will construct a directory in which intermediate data, intermediate results and final results are further organized in different subdirectories. Moreover, it also provides visualization tools for demonstrating final results as well as diagnosis plots from intermediate analysis steps. For more information and example data, the users can refer to the package website: http://zhangz05.u.hpc.mssm.edu/saasCNV/.

2 Materials

2.1 Computational Environment

The *saasCNV* package was implemented in R scripts [13] and can be run on desktop/laptop with commonly used operation systems (OS), such as Windows, Mac OS, and Linux. It can be also run on a Linux/Unix-based high performance cluster server, which is particularly efficient when processing multiple samples in parallel.

2.2 saasCNV Package

1. Install R package. The users can download OS specific, precompiled R package at https://cran.r-project.org/ and follow the step-by-step instructions for installation.
2. Install dependent packages. *saasCNV* uses functions in another two packages, *RANN* and *DNAcopy*, which needs to be installed a priori. The following commands are all typed and run in R environment following a prompt symbol ">", unless otherwise specified.

   To install RANN package, type

   ```
   > install.packages("RANN")
   ```

   To install *DNAcopy* package, type

   ```
   > source("https://bioconductor.org/biocLite.R")
   > biocLite("DNAcopy")
   ```
3. Install *saasCNV* package. You just need to type

   ```
   > install.packages("saasCNV")
   ```

Details about all functions in the package along with their parameters are provided in the software documents, which can be found by typing

```
> help("function.name")
```

where "*function.name*" refers to the name of the R function in the package to be searched.

2.3 Input Data

The analysis with *saasCNV* package starts with Variant Call Format (VCF) files generated from WGS or WES for the experimental design with tumor and matched normal pairs. The VCF files are usually produced by the GATK pipeline [14, 15] (*see* **Note 1**). The package can be also applied to data generated from whole-genome SNP array platforms. The analysis pipelines for WGS/WES data and SNP array data are almost identical except the data preparation step. We will mainly describe the pipeline for sequencing data analysis in Methods section, and explain the differences for SNP array data in **Note 2**.

To demonstrate how to use the package, we have prepared example data, which can be downloaded at https://zhangz05.u.hpc.mssm.edu/saasCNV/data/. The users can download these files: *WES_example.vcf.gz*, *vcf_table.txt.gz*, *snp_table.txt.gz*, *refGene_hg19.txt.gz* and *GC_1kb_hg19.txt.gz*; and put them in a working directory (e.g., *wk_dir*), where all analysis results and plots will be located.

3 Methods

The *saasCNV* package can be run in a "pipeline" mode with the integrated function *NGS.CNV* for sequencing data (*SNP.CNV* for SNP array data) or in a step-by-step mode. While the "pipeline" mode provides the users with a quick start, the step-by-step mode offers more control and flexibility. We refer the users to more details about "pipeline" mode in **Note 3**. Here we mainly describe the step-by-step mode, which illustrates the detailed workflow.

3.1 Input Data Preparation

At the beginning, you need to load *saasCNV* package in R environment and set up working directory:

```
> library(saasCNV)
> setwd("/path/to/wk_dir")
```

where "/path/to/wk_dir" specifies the absolute path to the working directory *wk_dir*. The pipeline for sequencing data (WGS/WES) analysis begins with VCF files. We have prepared an example WES VCF file *WES_example.vcf.gz*. The VCF file contains the information of both tumor and matched normal tissues.

Following the header of annotations, the first few rows are demonstrated in Table 1 (*see* more information in **Note 1**). The first nine columns in VCF file are mandatory, followed by the information of called variants from each sample starting from the tenth column. We provide a tool *vcf2txt* to retrieve necessary information from VCF file and convert it to a text table:

```
> vcf_table <- vcf2txt(vcf.file = "WES_example.
vcf.gz", normal.col = 9+1, tumor.col = 9+2)
```

The *normal.col* and *tumor.col* specify the columns in which the genotype and read depth information of normal and tumor tissues are respectively located in the VCF file. The first few lines of the resulting *vcf_table* are shown in Table 2. The first seven columns are retrieve from VCF file directly, where CHROM and POS are necessary for subsequent analysis. QUAL and MQ are quality scores for genotyping and reads mapping, which can be used as filters to exclude variants of poor quality (*see* **Note 1**). Starting from the eighth column are genotype, reference allele read depth, and alternative allele read depth for normal and tumor, respectively.

Then we transform allele-specific read depth information into log2ratio and log2mBAF [11] to be used for subsequent analysis using *cnv.data*:

```
> seq.data <- cnv.data(vcf = vcf_table, min.chr.
probe = 100, verbose = TRUE)
```

Here *min.chr.probe* specifies the minimum number of genotyped sites needed for a chromosome to be analyzed in subsequent steps. The first few rows of *seq.data* are shown in Table 3. The first four columns *chr*, *position*, *log2ratio*, and *log2mBAF* are the required information for SCNA analysis, while the rest columns are used for visualization and diagnosis. It should be noted that only heterozygous sites (i.e., *Normal.GT* is "0/1") in Table 2 were carried over to Table 3.

We provide an option to perform GC content adjustment on the log2ratio signal. The details can be found in **Note 4**.

3.2 Joint Segmentation

We employ a generalized circular binary segmentation (CBS) algorithm [16] to perform joint segmentation on log2ratio and log2mBAF dimensions. The function *joint.segmentation* outputs the starting and ending locations of each SCNA segment as well as several summary statistics.

```
> seq.segs <- joint.segmentation(data = seq.data,
min.snps = 10,
global.pval.cutoff = 1e-4, max.chpts = 30, verbose
= TRUE)
```

Table 1
An example VCF file

#CHROM	POS	ID	REF	Alt	Qual	Filter	Info	Format	Normal	Tumor
chr1	14,907	rs79585140	A	G	1650.44	VQSRTrancheSNP99.50to99.90	…; MQ = 20.64; …	GT:AD: …	0/1:42,43: …	0/1:56,46: …
chr1	14,930	rs75454623	A	G	2048.44	VQSRTrancheSNP99.50to99.90	…; MQ = 23.07; …	GT:AD: …	0/1:34,48: …	0/1:53,58: …
chr1	753,405	rs61770173	C	A	683.21	PASS	…; MQ = 29.16; …	GT:AD: …	1/1:0,24: …	1/1:0,11: …

The first few rows of the table are demonstrated. Some of the information that is not directly used in SCNA analysis is skipped and represented as ellipsis

Table 2
Allele-specific read depth information retrieved from the example VCF file

CHROM	POS	ID	REF	ALT	Qual	MQ	Normal. GT	Normal.REF.DP	Normal.ALT.DP	Tumor. GT	Tumor. REF.DP	Tumor. ALT.DP
chr1	762,589	rs71507461	G	C	898.2	37.9	1/1	2	19	1/1	0	14
chr1	762,592	rs71507462	C	G	880.2	37.9	1/1	2	19	1/1	0	14
chr1	762,601	rs71507463	T	C	831.2	37.45	1/1	1	20	1/1	0	12
chr1	762,632	rs61768173	T	A	618.23	37.39	1/1	1	16	1/1	0	8
chr1	801,943	rs7516866	C	T	1551.44	52.03	0/1	23	35	0/1	4	22
chr1	808,631	rs11240779	G	A	1173.37	54.69	0/1	19	13	1/1	3	24

The first few rows of the table are demonstrated

Table 3
Data used for joint segmentation and SCNA calling

chr	Position	log2ratio	log2mBAF	normal.BAF	normal.mBAF	tumor.BAF	tumor.mBAF
chr1	801,943	−1.550427058	0.487689879	0.603448276	0.603448276	0.846153846	0.846153846
chr1	808,631	−0.637998279	0.582147485	0.406250000	0.593750000	0.888888889	0.888888889
chr1	880,390	−0.463275109	0.610957709	0.523809524	0.523809524	0.800000000	0.800000000
chr1	881,627	−1.392885781	0.035334826	0.700000000	0.700000000	0.500000000	0.717356243
chr1	892,460	−0.039248827	0.783866569	0.444444444	0.555555556	0.956521739	0.956521739
chr1	898,852	−0.392885781	0.241008100	0.214285714	0.785714286	0.071428571	0.928571429

The first few rows of the table are demonstrated

Please refer to **Note 5** for details about the parameters used in this function. Table 4 lists the variables generated from the analysis along with their descriptions.

There is an option to merge adjacent segments, for which the median values in either or both dimensions of log2ratio and log2mBAF are not substantially different (*see* **Note 6**).

The results from joint segmentation (or segments merging) can be visualized chromosome by chromosome.

```
> diagnosis.seg.plot.chr(data = seq.data, segs =
seq.segs,
sample.id = "Sample ID", chr = 8)
```

Figure 2 shows the joint segmentation results for Chromosome 8 as an example (*see* **Note 5**).

3.3 SCNA Calling

In this step, we can assign SCNA status to each segment resulted from the joint segmentation step (or from the results after segments merging step). The baseline adjustment step is incorporated implicitly in the function *cnv.call*.

```
> seq.cnv <- cnv.call(data = seq.data, sample.id
= "Sample ID",
segs.stat = seq.segs, N = 1000, pvalue.cutoff =
0.05)
```

When this step is completed, a few more columns are added to *seq.segs*, which summarize the baseline adjusted log2ratio and log2mBAF mean/median values, p-values, and SCNA status for each segment (Table 4). The parameter *pvalue.cutoff* controls the significance level for SCNA calling (*see* **Note 7** for more information). We also provide an option to add gene annotation to each SCNA segment (*see* details in **Note 8**).

3.4 Visualization of Results

We provide two ways of visualization of joint segmentation and SCNA calling results: (1) the genome-wide SCNA profile; (2) cluster-type plot of SCNA segments projected onto the log2mBAF-log2ratio space.

```
> genome.wide.plot(data = seq.data, segs = seq.
cnv, sample.id = "Sample ID", chrs = 1:22)
> diagnosis.cluster.plot(segs = seq.cnv, chrs =
1:22, min.snps = 10)
```

The rendered figures are shown in Figs. 3 and 4. These plots are informative for diagnosis of data quality (*see* **Note 9** for more details) and manual baseline adjustment (*see* **Note 10** for more details).

Table 4
Variables generated from joint segmentation and SCNA calling

Variable	Description
Appear in both joint segmentation and SCNA calling results	
chr	Chromosome
posStart	The physical position of start site on the chromosome
posEnd	The physical position of end site on the chromosome
length	Physical size (posEnd - posStart +1)
chrIdxStart	The index of start site among all the sites on the chromosome
chrIdxEnd	The index of end site among all the sites on the chromosome
numProbe	The number of sites within the segment (chrIdxEnd - chrIdxStart +1)
log2ratio.Mean	The mean of log2ratio values at the sites within the segment
log2ratio.SD	The standard deviation of log2ratio values at the sites within the segment
log2ratio.Median	The median of log2ratio values at the sites within the segment
log2ratio.MAD	The median absolute deviation of log2ratio values at the sites within the segment
log2mBAF.Mean	The mean of log2mBAF values at the sites within the segment
log2mBAF.SD	The standard deviation of log2mBAF values at the sites within the segment
log2mBAF.Median	The median of log2mBAF values at the sites within the segment
log2mBAF.MAD	The median absolute deviation of log2mBAF values at the sites within the segment
Additional variables generated from SCNA calling results	
Sample_ID	Sample ID
remark	=1, if the identified baseline is not reliable; =0, otherwise
log2ratio.base.Mean	The baseline mean of log2ratio values at the sites within normal segments
log2ratio.base.Median	The baseline median of log2ratio values at the sites within normal segments
log2ratio.Sigma	The estimate noise level of log2ratio values across the genome
log2mBAF.base.Mean	The baseline mean of log2mBAF values at the sites within normal segments
log2mBAF.base.Median	The baseline median of log2mBAF values at the sites within normal segments
log2mBAF.Sigma	The estimate noise level of log2mBAF values across the genome
log2ratio.Mean.adj	The baseline-adjusted mean of log2ratio values at the sites within the segment
log2ratio.Median.adj	The baseline-adjusted median of log2ratio values at the sites within the segment

(continued)

Table 4 (continued)

Variable	Description
log2mBAF.Mean.adj	The baseline-adjusted mean of log2mBAF values at the sites within the segment
log2mBAF.Median.adj	The baseline-adjusted median of log2mBAF values at the sites within the segment
log2ratio.p.value	The p-value for the log2ratio.Median.adj value of the segment
log2mBAF.p.value	The p-value for the log2mBAF.Median.adj value of the segment
p.value	The p-value for both log2ratio.Median.adj and log2mBAF.Median.adj values of the segment
CNV	SCNA status, including loss, gain, CN-LOH, normal, undecided

4 Notes

1. Detailed description of VCF can be found at https://samtools.github.io/hts-specs/VCFv4.2.pdf. In GATK pipeline, the detected variants are subject to variant quality score recalibration (VQSR) so that the FILTER field in generated VCF files will be populated with phrases, such as "PASS", "VQSRTrancheSNP99.50to99.90", and so on, which describe the quality of the variant calls (Table 1). The *vcf2txt* function uses the variant sites with the quality of "PASS" for downstream SCNA analysis. When the FILTER field of the input VCF files is not filled with such quality-descriptive values or the VCF files are generated by other tools (e.g., MuTect [17], VarScan2 [18], etc.), the users may need to modify *vcf2txt* function to make it adaptable for specific data format.
2. We mainly consider the SNP array data produced by Illumina Infinium whole-genome microarray. For SCNA analysis, the log R ratio (LRR) and B allele frequency (BAF) signals, which respectively reflect the total copy number change and allele proportion change [8], can be retrieved from final report files (*see* Table 5 for an example). It should be noted that, in Table 5, the columns ID, REF, and ALT are not essential and only for annotation purpose. For data generated from Affymetrix SNP array, LRR and BAF information can be extracted by PennCNV-Affy tools (http://penncnv.openbioinformatics.org/en/latest/user-guide/affy/).
3. In the "pipeline" mode, all the analysis steps described in Methods section are integrated into *NGS.CNV* function and can be run altogether. The results, including tables and plots,

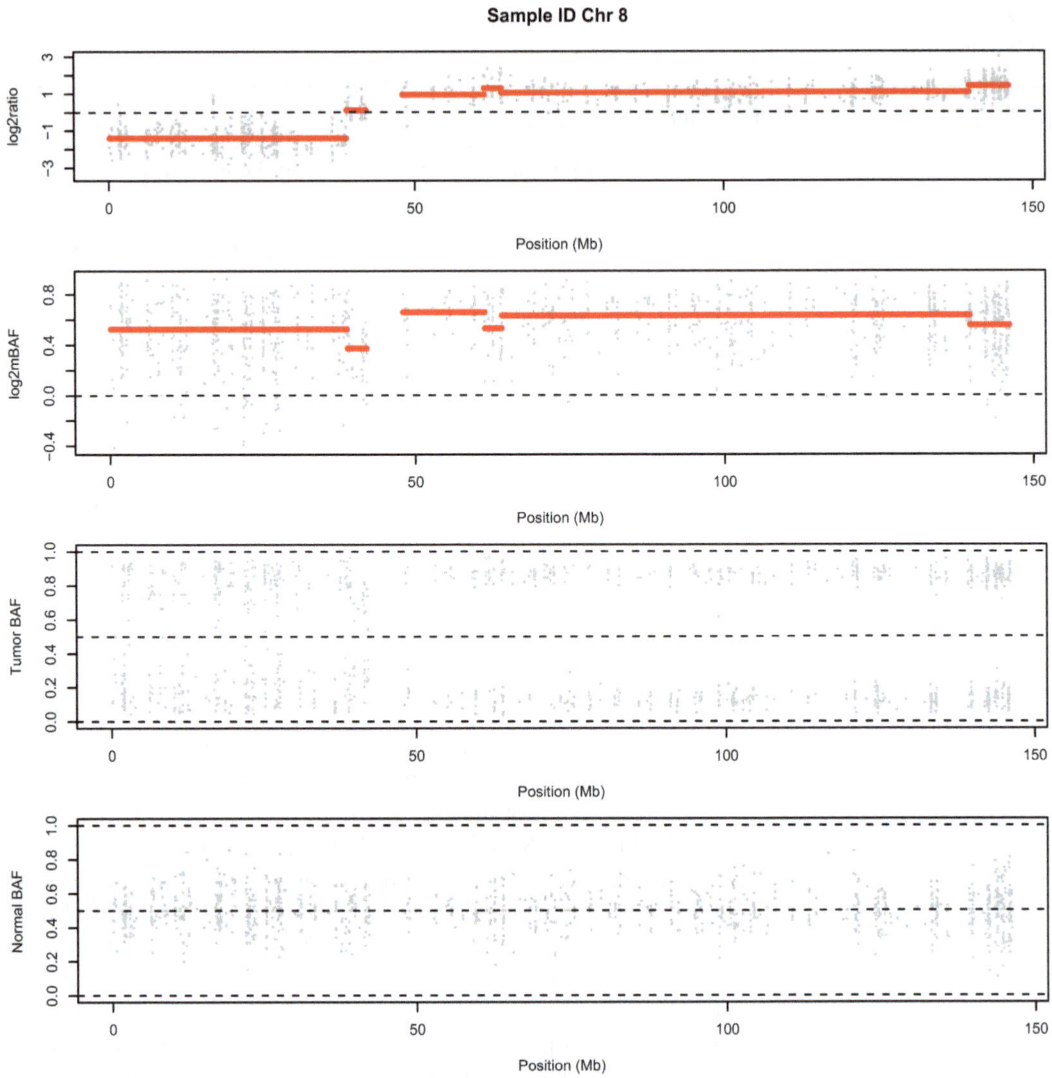

Fig. 2 Visualization of joint segmentation results on a chromosome. From top to bottom, the panels display the signals of log2ratio, log2mBAF, tumor BAF, and normal BAF, respectively. The gray dots indicate the data points measured at genotyped loci ordered by their physical locations across the chromosome. The red segments indicate the SCNA segments produced by the joint segmentation algorithm

are placed in subdirectories of the working directory *output.dir* specified by the users.

```
> vcf_table <- read.delim(file = "vcf_table.txt.
gz", as.is = TRUE)
> NGS.CNV(vcf = vcf_table, output.dir = "/path/
to/wk_dir",
sample.id = "Sample ID", min.chr.probe = 100,
do.GC.adjust = FALSE, min.snps = 10, joint.seg-
mentation.pvalue.cutoff = 1e-4, max.chpts = 30,
```

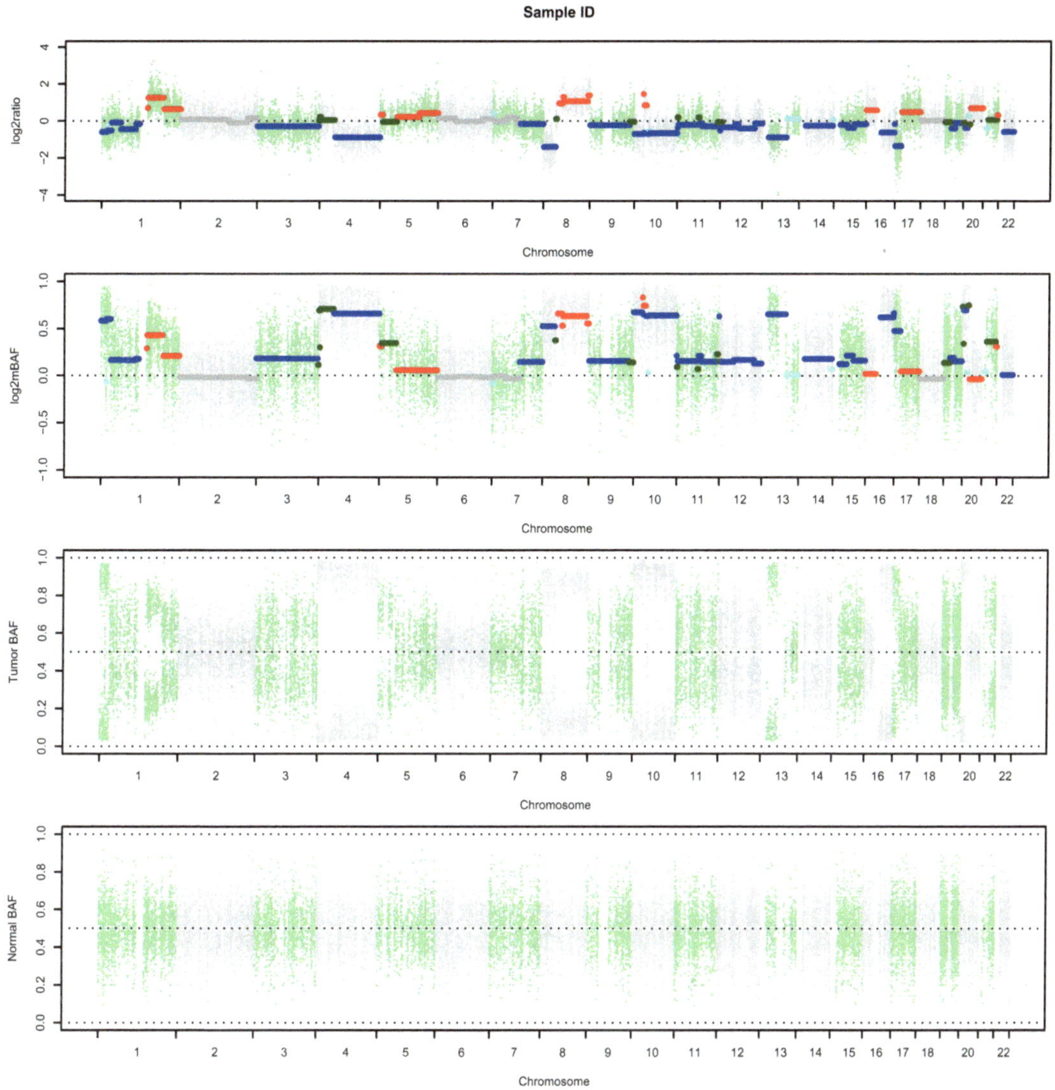

Fig. 3 Genome-wide visualization of SCNA profile. From top to bottom, the panels display the signals of log2ratio, log2mBAF, tumor BAF, and normal BAF, respectively. The dots, each representing a genotyped locus, are ordered by their chromosomal locations and colored alternately to distinguish chromosomes. The segments, each representing a DNA segment resulting from the joint segmentation algorithm, are colored based on inferred SCNA status: blue—loss; gray—normal; red—gain; green—CN-LOH; cyan—undecided

```
do.merge = FALSE, num.perm = 1000, do.cnvcall.
on.merge = FALSE,
cnvcall.pvalue.cutoff = 0.05, do.plot = TRUE,
do.gene.anno = TRUE, gene.anno.file = "refGene_hg19.
txt.gz", seed = 123456789, verbose = TRUE)
```

The *seed* parameter is specified to ensure reproducibility of the results. The "pipeline" mode for SNP array data is imple-

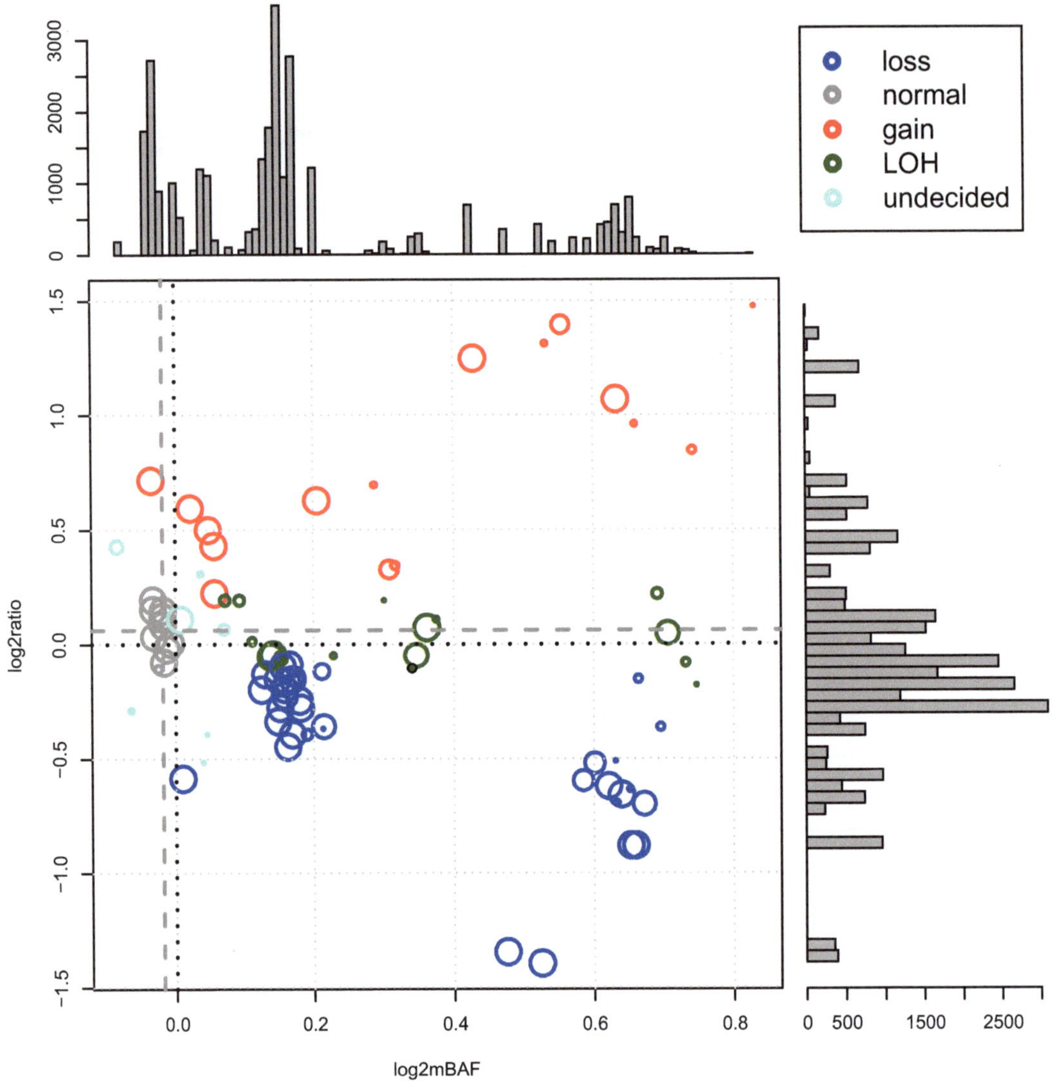

Fig. 4 Cluster-type plot of SCNA profile. On the main log2mBAF-log2ratio panel, each circle corresponds to a segment in Fig. 3, with the size reflecting the length of the segment; the color code is specified in legend; the dashed gray lines indicate the adjusted baselines. The side panels, corresponding to log2ratio and log2mBAF dimension respectively, show the distribution of the median values of each segment

mented as the function *SNP.CNV* along with similar parameters as *NGS.CNV*.

4. When the tumor–normal pair experiment design is properly carried out, the spatial variability in log2ratio signal, due to nonuniform GC content distribution and other factors, can be effectively neutralized by normalizing tumor data with matched normal data [11]. In most cases, this step is not necessary. We provide an optional function GC.adjust to adjust for GC content when the log2ratio variability from GC content is not

Table 5
An example of information retrieved from SNP array data for SCNA analysis

CHROM	POS	ID	REF	ALT	Normal. GT	Normal. LRR	Normal. BAF	Tumor. GT	Tumor. LRR	Tumor. BAF
chr1	768,448	rs12562034	G	A	0/1	−0.2072	0.6340	0/0	−0.3481	0.9913
chr1	1,005,806	rs3934834	C	T	0/0	0.2237	1.0000	0/0	−0.4176	1.0000
chr1	1,018,704	rs9442372	G	A	0/0	−0.3591	1.0000	0/0	−0.6717	1.0000
chr1	1,021,415	rs3737728	C	T	0/0	−0.2973	1.0000	0/0	−0.8547	0.9581
chr1	1,021,695	rs9442398	G	A	0/0	0.4503	0.9993	0/0	−0.3396	0.9988
chr1	1,030,565	rs6687776	C	T	0/0	0.1752	1.0000	0/0	−0.2328	0.9983

The first few rows of the table are demonstrated

fully neutralized by normal data. Along with the function, we also provide an example GC_1kb_hg19.txt.gz file, which summarizes GC content in 1 kb window (hg19 assembly) for GC adjustment.

```
> gc <- read.delim(file = gc.file, as.is=TRUE)
> seq.data <- GC.adjust(data = seq.data, gc = gc)
```

After GC content adjustment, the GC-adjusted values will replace the original values in the *log2ratio* column of *seq.data*, while the original values will be stored in an additional column called *log2ratio.woGCAdj*. The updated *seq.data* can be directly fed to the subsequent steps.

5. In the *joint.segmentation* function, the parameter *min.snps* specifies the minimum number of genotyped loci across which a segment is required to span; *global.pval.cutoff* is the p-value cut-off for change points to be determined as significant in each cycle of joint segmentation; *max.chpts* controls the maximum number of change points to be detected for each chromosome—the recursive segmentation procedure will halt when reaching this limit. The combinations of these parameters together control the sensitivity of the joint segmentation step. Figure 2 shows an example of diagnosis plot for the results generated from joint segmentation. By inspecting the goodness-of-fit of the red bars to data points, the users can have an idea whether the parameters specified in the *joint.segmentation* function are too stringent or not, and then adjust them when necessary.
6. For WGS and SNP array data, merging step is recommended, whereas for WES data, this step is not necessary. To perform merging step, the users can use the *merging.segments* function.

```
> seq.segs.merge <- merging.segments(data = seq.
data, segs.stat = seq.segs,
N = 1000, merge.pvalue.cutoff = 0.05, verbose=TRUE)
```

The parameter *N* is the number of bootstrap samples to be drawn for estimation of the p-value and *merge.pvalue.cutoff* controls the significance level at which two adjacent segments are considered significantly apart—if the *p*-value is above the cutoff, the two adjacent segments will be merged back to one piece. The output *seq.segs.merge* looks similar to the table *seq.segs* produced from the joint segmentation step, except that some of the rows from joint segmentation may be merged.

7. Regarding the choice of *pvalue.cutoff* in the *cnv.call* function, the study [11] provided useful guidance. When the *pvalue.cut-*

off varies from 0.001 to 0.05, the sensitivity and specificity of SCNA calls are rather stable, ranging around 90%, with smaller p-value favoring relatively higher specificity and lower sensitivity and vice versa. In practice, the users can choose *pvalue.cutoff* within the range from 0.001 to 0.05 depending on their preference for higher sensitivity or specificity.

8. The RefSeq gene annotation file can be downloaded from UCSC Genome Browser (https://genome.ucsc.edu/). We prepared a file *refGene_hg19.txt.gz* with gene annotation information, such as gene symbol and start/end coordinates (hg19 assembly).

```
  > gene.anno <- read.delim(file = "refGene_hg19.
txt.gz", as.is = TRUE, comment.char = "")
  > seq.cnv.anno <- reannotate.CNV.res(res = seq.
cnv, gene = gene.anno, only.CNV = TRUE)
```

   By turning on the *only.CNV* flag, only those segments called as loss, gain, or CN-LOH will be annotated and output to the resulting table, in which two additional columns are generated, *numGene* and *gene*, indicating the number of genes and the gene symbols within each segment.

9. In Fig. 3, the genome-wide SCNA profile is displayed in different perspectives: log2ratio, log2mBAF, tumor BAF, and normal BAF, respectively. From this plot, the users can have a quick overview of the whole cancer genome and qualitative answers to questions like: (1) Is the genome perturbed by SCNA dramatically? (2) Which chromosome arm is alternated by broad events? (3) Is there any focal event? From the signal patterns of the bottom two panels, the users can confirm whether the identity of tumor and normal tissue is correct—tumor BAF signal exhibits several symmetric bands of data points deviating from 0.5, whereas normal BAF signal fluctuates around 0.5 with random noise. Mislabeling of tumor and normal samples, if there is any, can be quickly identified by such diagnosis plot. In Fig. 4, each segment from Fig. 3 is represented by a circle in the log2mBAF-log2ration space, color-coded based on inferred SCNA status. Segments with the same SCNA status are clustered together. This is a good way to diagnose (1) whether the baseline cluster of normal segments (the cluster of gray circles in Fig. 4) is correctly identified, and (2) whether the SCNA status (loss/gain/CN-LOH) is correctly assigned to each segment.

10. In the *cnv.call* function, we incorporate a feature that facilitates the users to manually adjust the baseline when automatic adjustment is not correct in the first running of *cnv.call*, based on a visual inspection of the diagnosis plot (e.g., Fig. 4). The

users can manually adjust the baseline by specifying the boundaries within which the "normal" cluster is located and make the SCNA call (run *cnv.call*) again.

```
> seq.cnv <- cnv.call(data = seq.data, sample.id
= "Sample ID",
segs.stat = seq.segs, N = 1000, pvalue.cutoff =
0.05,
do.manual.baseline = TRUE, log2mBAF.left = -0.05,
log2mBAF.right = 0.05, log2ratio.bottom = -0.15,
log2ratio.up = 0.15)
```

Here the parameter *do.manual.baseline* is the switch for activating the manual adjustment function; *log2mBAF.left*, *log2mBAF.right*, *log2ratio.bottom*, and log2ratio.up are used to manually specify the four boundaries of "normal" cluster. A diagnosis plot such as Fig. 4 can provide guidance on locating the boundaries.

References

1. Zack TI, Schumacher SE, Carter SL et al (2013) Pan-cancer patterns of somatic copy number alteration. Nat Genet 45(10):1134–1140. https://doi.org/10.1038/ng.2760
2. Kim TM, Xi R, Luquette LJ et al (2013) Functional genomic analysis of chromosomal aberrations in a compendium of 8000 cancer genomes. Genome Res 23(2):217–227. https://doi.org/10.1101/gr.140301.112
3. Wang K, Lim HY, Shi S et al (2013) Genomic landscape of copy number aberrations enables the identification of oncogenic drivers in hepatocellular carcinoma. Hepatology 58(2):706–717. https://doi.org/10.1002/hep.26402
4. Torrecilla S, Sia D, Harrington AN et al (2017) Trunk mutational events present minimal intra- and inter-tumoral heterogeneity in hepatocellular carcinoma. J Hepatol 67(6):1222–1231. https://doi.org/10.1016/j.jhep.2017.08.013
5. Uzilov AV, Ding W, Fink MY et al (2016) Development and clinical application of an integrative genomic approach to personalized cancer therapy. Genome Med 8(1):62. https://doi.org/10.1186/s13073-016-0313-0
6. Gusnanto A, Wood HM, Pawitan Y et al (2012) Correcting for cancer genome size and tumour cell content enables better estimation of copy number alterations from next-generation sequence data. Bioinformatics 28(1):40–47. https://doi.org/10.1093/bioinformatics/btr593
7. Pinkel D, Segraves R, Sudar D et al (1998) High resolution analysis of DNA copy number variation using comparative genomic hybridization to microarrays. Nat Genet 20(2):207–211. https://doi.org/10.1038/2524
8. Peiffer DA, Le JM, Steemers FJ et al (2006) High-resolution genomic profiling of chromosomal aberrations using Infinium whole-genome genotyping. Genome Res 16(9):1136–1148. https://doi.org/10.1101/gr.5402306
9. Campbell PJ, Stephens PJ, Pleasance ED et al (2008) Identification of somatically acquired rearrangements in cancer using genome-wide massively parallel paired-end sequencing. Nat Genet 40(6):722–729. https://doi.org/10.1038/ng.128
10. Chiang DY, Getz G, Jaffe DB et al (2009) High-resolution mapping of copy-number alterations with massively parallel sequencing. Nat Methods 6(1):99–103. https://doi.org/10.1038/nmeth.1276
11. Zhang Z, Hao K (2015) SAAS-CNV: a joint segmentation approach on aggregated and allele specific signals for the identification of somatic copy number alterations with next-generation sequencing data. PLoS Comput Biol 11(11):e1004618. https://doi.org/10.1371/journal.pcbi.1004618
12. Kim HY, Choi JW, Lee JY et al (2017) Gene-based comparative analysis of tools for estimating copy number alterations using whole-exome sequencing data. Oncotarget 8(16):27277–27285. https://doi.org/10.18632/oncotarget.15932
13. Core Team R (2016) R: A language and environment for statistical computing. R Foundation for Statistical Computing, Vienna

14. McKenna A, Hanna M, Banks E et al (2010) The genome analysis toolkit: a MapReduce framework for analyzing next-generation DNA sequencing data. Genome Res 20(9):1297–1303. https://doi.org/10.1101/gr.107524.110
15. DePristo MA, Banks E, Poplin R et al (2011) A framework for variation discovery and genotyping using next-generation DNA sequencing data. Nat Genet 43(5):491–498. https://doi.org/10.1038/ng.806
16. Zhang NR, Siegmund DO, Ji H et al (2010) Detecting simultaneous changepoints in multiple sequences. Biometrika 97(3):631–645. https://doi.org/10.1093/biomet/asq025
17. Cibulskis K, Lawrence MS, Carter SL et al (2013) Sensitive detection of somatic point mutations in impure and heterogeneous cancer samples. Nat Biotechnol 31(3):213–219. https://doi.org/10.1038/nbt.2514
18. Koboldt DC, Zhang Q, Larson DE et al (2012) VarScan 2: somatic mutation and copy number alteration discovery in cancer by exome sequencing. Genome Res 22(3):568–576. https://doi.org/10.1101/gr.129684.111

Chapter 3

Statistical Detection of Genome Differences Based on CNV Segments

Yang Zhou, Derek M. Bickhart, and George E. Liu

Abstract

Population analysis using copy number variation (CNV) is far more complex than analysis using SNPs because of the diverse copy number and inconsistent boundaries of CNVs in different individuals that causes changes in frequency. Multiple studies have reported CNV regions associated with diseases or body traits based on a CNV segmentation strategy that condenses calls from multiple different sources into a genotype state. Here, we provide a guideline of how to generate CNV segments from known CNV results, and how to detect genome differences based on CNV segments.

Key words CNV segments, PennCNV, F_{ST}, Lineage differences, Misassemblies

1 Introduction

Unlike single nucleotide variation, the unique properties of CNVs cause difficulties in the comparison of CNVs between two or more populations directly. The copy number of CNV regions in the genome can range from a copy number of 0 (homozygous deletion) to more than 3 (duplication) that produces multiple genotypes (e.g., 0 copy, 1 copy, 2 copies, 3 copies, ...) [1]. Moreover, many CNV calls are detected with different boundaries in different individuals [2]. These CNVs overlap with each other but have different break points in different individuals—this makes it difficult to calculate genotype frequencies for CNVs as one of the overlapping events could have arisen independently of the others. Due to this, the application of CNV analysis in population genetics has long been held back and most CNV studies have been limited to CNV detection and CNV region (CNVR) description rather than statistical analysis.

CNV segments are defined as partial regions of the CNV that enable comparisons of CNVs between populations [1, 3, 4]. CNV segments with consistent break points among different individuals can be genotyped and a genotype frequency can be calculated for

Derek M. Bickhart (ed.), *Copy Number Variants: Methods and Protocols*, Methods in Molecular Biology, vol. 1833, https://doi.org/10.1007/978-1-4939-8666-8_3, © Springer Science+Business Media, LLC, part of Springer Nature 2018

the event. Several programs, including CNVruler and ParseCNV, use algorithms based on CNV segments to perform case–control genome-wide association study (GWAS) analysis [5, 6]. Golden Helix SNP & Variation Suite (SVS) software applies a method to detect common CNV segments using a powerful segmenting algorithm (Copy Number Analysis Method, CNAM) on a multisample (multivariate) basis that can be further used to perform GWAS analysis [7]. CNV segment analysis is also useful as a follow-up in the use of CNV detection software, such as PennCNV, as it can be used to detect population signature selection by normalizing detected CNV into CNV segments [8]. Application of CNV segment analysis has helped to identify multiple regions and genes that may play roles in diseases, body traits, and breed differentiation [8–10].

SNP array CNV calling makes use of several common metrics gained from SNP genotyping analysis. The LogR ratio (LRR) from SNP arrays is used to classify each CNV as loss, neutral, and gain in the SNP microarray data [11]. By combining the B allele frequency (BAF) information, software such as PennCNV and cnvPartion are able to detect copy numbers of CNV including 0 copy, 1 copy, 2 copies, 3 copies, and 4 or more copies [7, 12]. In this chapter, we describe methods to generate CNV segments based on PennCNV results. These results are further used to detect genome signature selection between populations and genome misassemblies by performing F_{ST} statistics and probe distribution analyses. This will guide the readers to flexibly handle existing CNV results or newly called CNV regions and further detect genome differences by involving more detail genotype categories and genotype frequencies.

2 Materials

The material used in this chapter is the CNV call results from at least two populations. Multiple software can call CNVs from SNP microarray data, including cnvPartition, Birdsuite, PennCNV, QuantiSNP, and so on [7]. They may produce results with diverse formats; however, the final result file will include the CNV coordinate information on the chromosomes for each sample. Here, we use PennCNV results as an example. The column names of the PennCNV output are the following: Chromosome region, Number of SNP supports the CNV region, Length, Status and copy number, Sample name, Start SNP, End SNP. Readers are referred to the extensive PennCNV website for more information (http://penncnv.openbioinformatics.org/en/latest/user-guide/test/).

3 Method

3.1 Generate CNV Segments from Known CNVs; an Example from PennCNV

Traditionally, CNV calls from different software tools will be merged to CNVRs (CNV regions) followed by statistic of the CNVR size distribution and status without considering the frequencies of the CNV located in the CNVRs (Fig. 1a). By contrast, in this chapter, we will first normalize CNV calls into CNV segments according to their base pair boundaries (Fig. 1b).

The sample data used in this chapter can be downloaded from the publisher's website (http://extras.springer.com/). PennCNV results derived from BovineHD Genotyping BeadChip Data of two cattle populations (882 unrelated *Bos indicus* individuals and 316 highly diverse individuals of 12 *Bos taurus* breeds, *see* **Note 1**) will be used as the primary data.

3.1.1 Prepare the CNV Information

To get consistent CNV segments, the first step is to merge the PennCNV results of the two populations. The first column with the CNV coordinate information will be used to generate a bed format including the chromosome, start and end of the unique CNVs.

```
indicus<-read.table("penncnv_indicus.txt")
taurus<-read.table("penncnv_taurus.txt")
#merge the PennCNV results of indicus and taurus
cattle.
mydata<-rbind(indicus,taurus)
data 1<-strsplit(as.character(unique(myda
ta[,1])),":")
data 2<-do.call(rbind,data 1)
data 3<-strsplit(data 2[,2],"-")
data 4<-do.call(rbind,data 3)
data 5<-cbind(data 2[,1],data 4)
write.table(data 5,"cnv.bed",sep="\
t",quote=FALSE,row.names=FALSE,col.names=FALSE)
```

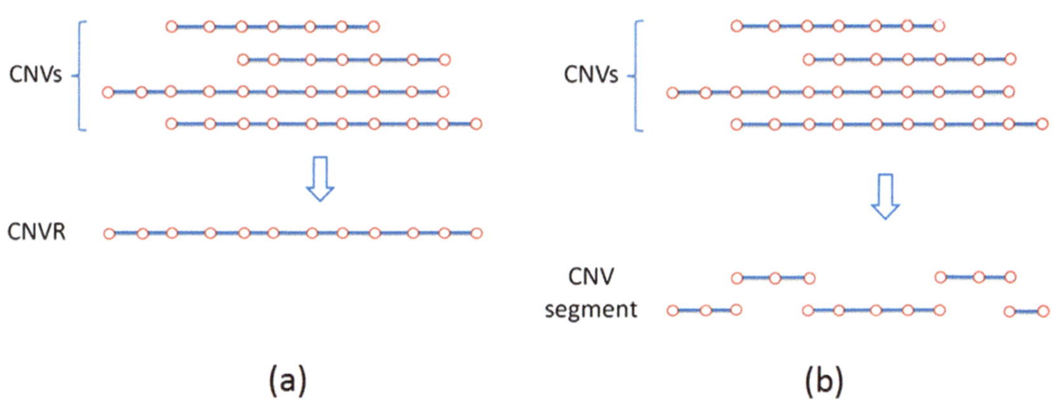

Fig. 1 The strategies to merge CNVs to CNVRs and to disjoin CNVs to CNV segments. (**a**) The strategy to merge CNVs to CNVRs. (**b**) The strategy to disjoin CNVs to CNV segments. The red circles represent the SNPs used to call CNVs

3.1.2 Normalize CNVs into CNV Segments

We supply two methods for normalization of the CNVs into CNV segments. One can also perform this step using an in-house script.

1. R packages (GenomicRanges, https://bioconductor.org/packages/release/bioc/html/GenomicRanges.html) are used for the normalization. But please note that it may produce small artificial CNV segments (*see* **Note 2**). The code for loading data into the R package is listed below:

```
library(GenomicRanges)
CNV<-data 5
start<-as.numeric(CNV[,2])
end<-as.numeric(CNV[,3])
CNV_GRanges<-GRanges(seqnames=CNV[,1],
IRanges(start,end))
CNV_segment<-disjoin(CNV_GRanges)
#Get the raw CNV segments
raw_CNV_segment<-data.frame(seqnames=seqnames(CNV_
segment),start=start(CNV_segment),end=end(CNV_seg-
ment))
#Filter the CNV segments smaller than 50 bp (see
Note 2)
good_ CNV_segment <- raw_CNV_segment[which(raw_CNV_
segment[,3]- raw_CNV_segment[,2] >= 50),]
write.table(good_ CNV_segment," good_ CNV_segment.
bed",sep="\t",quote=FALSE,row.names=FALSE,col.
names=FALSE)
```

2. One can also use bedtools to generate the CNV segments using the command line as following.

```
bedtools genomecov -i cnv.bed -g genome.txt > cnv_
seg_temp
awk -v OFS='\t' "$4-$3>=50 {print chr$2, $3, $4}
cnv_seg_temp" > good_ CNV_segment_bed
```

The good_ CNV_segment_bed file contains the CNV segment coordinate information (chromosome, start and end) needed for further analysis.

3.2 Detect Signature Selection Using F_{ST} Based on CNV Segments

F_{ST} is one of the most commonly used methods for detecting signatures of selection in SNP studies since it is directly related to the variance in allele frequency between populations. After we generate CNV segments, we can detect signature selection using F_{ST} to find out the genome differences between two or more populations.

3.2.1 Genotype the CNV Segments

In the PennCNV results, the CNVs are defined as five different statuses with 0, 1, 3, ≥4 and normal (2 for diploid organisms) copies. The CNV segments have same statues with the CNVs that they originate from and are genotyped as 0, 1, 2, 3, 4 for different individuals according to the copy number. The number of animals for each genotype can be summarized according to the following R code.

```
population<-c("indicus","taurus")
for(i in 1:length(population)){
  data<-read.table(paste("penncnv_",population
[i],".txt",sep=""))
  data1<-strsplit(as.character(data[,1]),":")
  data2<-do.call(rbind,data1)
  data3<-strsplit(data2[,2],"-")
  data4<-do.call(rbind,data3)
  data5<-cbind(data2[,1],data4)
  data6<-data.frame(cbind(data5[,1],data5[,2],data5
  [,3],data[,4]))
  animal_number<-length(unique(data[,5]))
  genotype_number<-NULL
  for(j in 1:dim(good_CNV_segment)[1]){
     targetCNV<-
     data6[which(data6[,1]==good_CNV_segment[j,1]&
     as.numeric(data6[,2])<=good_CNV_segment[j,2]&
     as.numeric(data6[,3])>=good_CNV_segment[j,3]),]
     abnormal_number<-dim(targetCNV)[1]
     CNV_0<-targetCNV[which(targetCNV[,4]=="0"),]
     CNV_1<-targetCNV[which(targetCNV[,4]=="1"),]
     CNV_3<-targetCNV[which(targetCNV[,4]=="3"),]
     CNV_4<-targetCNV[which(targetCNV[,4]=="4"),]
     copy_number_0<-dim(CNV_0)[1]
     copy_number_1<-dim(CNV_1)[1]
     copy_number_3<-dim(CNV_3)[1]
     copy_number_4<-dim(CNV_4)[1]
     copy_number_2<-animal_number-abnormal_number
     CNV_segment_genotype_number<-cbind(copy_
     number_0,copy_number_1,copy_number_2,copy_
     number_3,copy_number_4)
     genotype_number<-rbind(genotype_number,CNV_
     segment_genotype_number)
}
all_CNV_segment_genotype_number<-cbind(good_CNV_
segment,genotype_number)
write.table(all_CNV_segment_genotype_
number,paste("all_CNV_segment_genotype_number_",pop
ulation[i],sep=""),sep="\t",quote=FALSE)
}
```

3.2.2 Calculate the F_{ST} Values Between the Two Populations Based on CNV Segments

Next, the frequency of the CNV genotypes can be calculated and used for F_{ST} analysis. Here is the formula to calculate the F_{ST} value.

$$F_{ST} = (H_t - H_s) / H_t.$$

$$H_t = 1 - \mathrm{P}\, t_i^2;$$

$$t_i = ((x_i \cdot N_x) + (y_i \cdot N_y)) / (N_x + N_y).$$

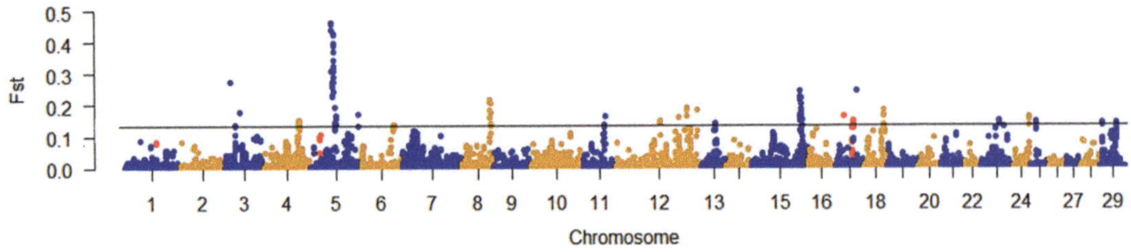

Fig. 2 Manhattan plot for the F_{ST} between indicus and taurine cattle at CNV segment level. The red dots represent false positive CNV segments which are representative of reference genome misassemblies (for more information, please see Subheading 3.4)

$$H_s = \left(\left(1 - P_{xi}^{\ 2}\right) \cdot N_x + \left(1 - P_{yi}^{\ 2}\right) \cdot N_y\right) / \left(N_x + N_y\right).$$

where x_i and y_i are the population frequencies of allelic CNV segment number i ($i = A_0, A_1, A_2, A_3, A_4$ or $> A_4$) in population X and Y, respectively, N_x and N_y denote the number of individuals in population X and Y, and t_i is a weighted average of x_i and y_i.

Following the supplied formula for F_{ST}, we can calculate the F_{ST} value for each CNV segment between the *Bos taurus* and *Bos indicus* populations. We used a cutoff threshold of $F_{ST} \geq 0.14$ (0.5% tail) to evaluate the differentiation. The result can be quickly visualized using a Manhattan plot (Fig. 2). The Manhattan plot helps to highlight the significantly different CNV segments.

```
library(qqman)
manhattan(inputfile,chr="chr",bp="position",p="Fst",
snp="SNP",logp=FALSE,ylim=c(0,0.5),col = c("blue4",
"orange3"), ylab="Fst") (see Note 3)
```

3.2.3 Explore Genes in the Selection Regions

Significantly similar CNV segments will be merged as CNVRs according to their adjacent relationship (*see* **Note 4**). After defining the signature selection regions, a genome browser or custom scripts can be used to identify genes located in or overlapped with the selection regions that may function in the divergence between populations (*see* **Note 5**). To get more specific functions for those genes, we can check the gene functions according to previous published papers, gene databases for individual gene, or apply gene ontology (GO) analysis for all genes (*see* **Note 6**).

3.3 Visualization of the LRR and BAF in the Significantly Different CNV Segments

PennCNV detects the CNV regions according to the total signal intensity (LRR), the B-allele frequency (BAF), and other information at each SNP marker [12]. To get more detailed information and to validate the results, we can zoom in the signature selection regions merged from different CNV segments to check the LRR and BAF distributions. One signature selection region on Chr5 can be used as an example. It is a loss type CNV of a set of differentiated segments occurring in 66.33~79.93% of the indicus cattle while being present in only 4.75–6.32% of taurine cattle. The LRR and

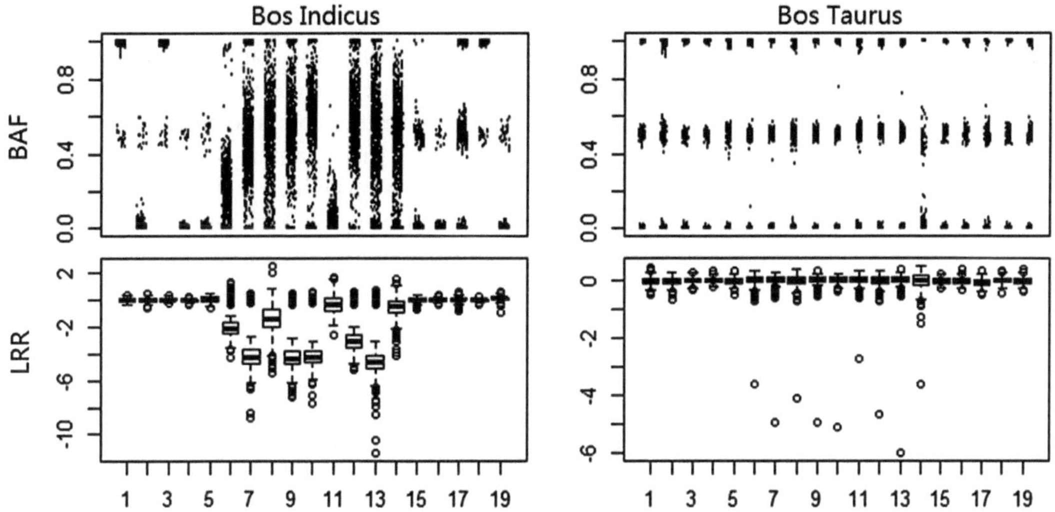

Fig. 3 BAF and LRR distributions of the SNPs in the signature selection region. Five upstream and downstream SNPs were included to represent the basal level

BAF information for each SNP in the corresponding and adjacent region can be extracted and prepared for plotting (*see* **Note 7**).

```
indicus_BAF<-read.table("indicus_BAF.txt",header=T)
taurine_BAF<-read.table("taurine_LRR.txt",header=T)
indicus_LRR<-read.table("indicus_LRR.txt",header=T)
taurine_LRR<-read.table("taurine_LRR.txt",header=T)
par(mfrow=c(2,2),oma=c(2,2,0,0),mar=c(1,2,1,2))
plot(indicus_BAF$position,indicus_BAF$BAF,cex =
.3,xaxt="n")
plot(taurine_BAF$"position",taurine_BAF$"BAF",cex =
.3,xaxt="n")
boxplot(indicus_LRR $Values~ indicus_LRR$Probes
,Xlab="Probes", Ylab="LogR ratio")
boxplot(taurine_LRR $Values~ taurine_LRR $Probes,
Xlab="Probes",Ylab="LogR ratio")
```

One can also refer to the PennCNV user manual to plot the LRR and BAF distributions (http://penncnv.openbioinformatics.org/en/latest/user-guide/visualization/).

In Fig. 3, except for the LRR and BAF of six probes in the signature selection region, an additional five probes upstream and downstream of the CNV segment are extracted to serve as controls. In both indicine and taurine cattle, the LRRs and BAFs of the ten control SNPs shows normal distribution with the LRRs around 0 and BAFs clustered as genotypes with two alleles. In the signature selection region, the average LRRs are lower in indicus cattle than that of taurine cattle and BAFs are dispersedly distributed in indicus cattle. In taurine cattle, LRR and BAF values are clearly clustered as genotypes with two alleles, which implies the deletion may be a true positive event and is enriched in indicus cattle (*see* **Note 8**).

3.4 Detect Misassemblies Within ChrY and Autosomes

Misassemblies (or errors in the reference genome) within the sex chromosomes and autosomes will cause artificial CNVs when female and male animal genotypes are used to call CNVs. For example, if the misassemblies result from the insertion of ChrY sequence into autosomes, the female animals will be called with deletion in the corresponding regions which will not be seen in the male animals. In cattle, sequence in the Y chromosome is potentially misassembled to autosomes that cause artificial CNVs (*see* **Note 9**). We can detect those regions using similar steps in Subheadings 3.3 and 3.4 by comparing the female group and male group in the *Bos taurus* and *Bos indicus* populations.

We use *Bos indicus* population as an example. In the Manhattan plot (Fig. 4a), over 99% CNV segments had F_{ST} value lower than 0.1 and some CNV segments are with abnormally high F_{ST} values on Chr1, Chr5, and Chr17. One region ranging from 25,056,695 to 25,119,996 in Chr17 is further visualized in terms of the LRRs and BAFs (Fig. 4b). In female animals, all SNPs are with low LRR and abnormal BAF patterns as compared to those in male samples. Additionally, there was a lack of heterozygous calls for male cattle in the SNPs, which shows only one allele existed in each male sample (*see* **Note 10**). This implies that the region in the reference genome does not exist in the female animals while only exist one copy in the male animals. There should be no differences in the

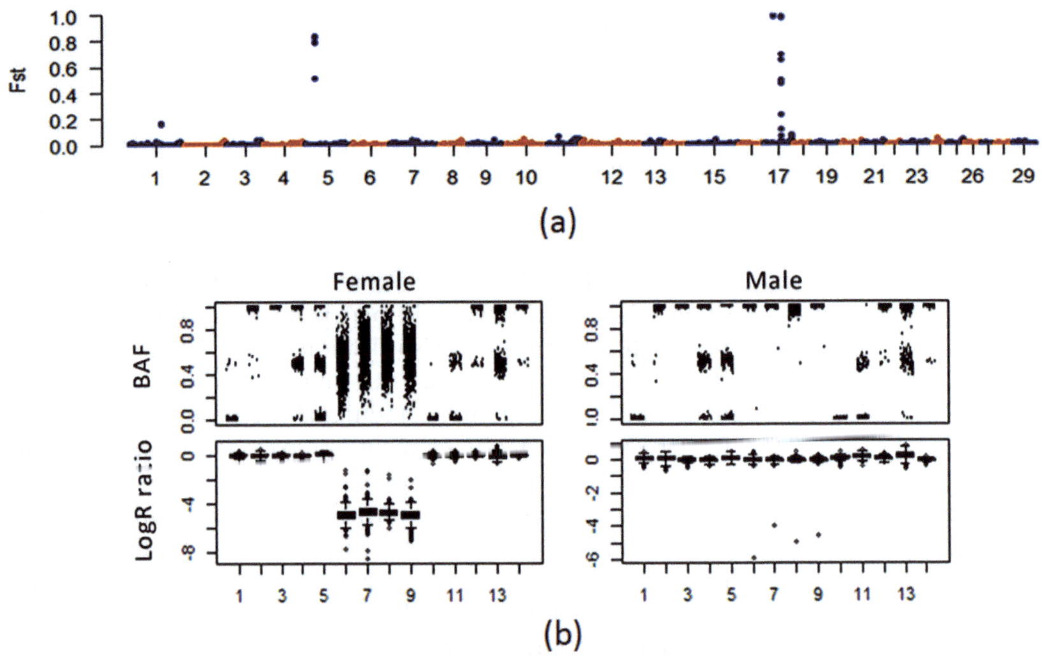

Fig. 4 Detect misassemblies between ChrY and autosomes. (**a**) Manhattan plot for the F_{ST} between female and male cattle at CNV segment level. (**b**) BAF and LRR distributions of the SNPs in the signature selection region

autosome sequences between male and female animals. Thus the different CNV regions should come from ChrY. The result can be further confirmed by male-specific PCR and fluorescence in situ hybridization (FISH).

4 Notes

1. To minimize the effects of close relationships and genome sharing among samples, the highly related individuals with pi-hat (an identity-by-descent or IBD estimation) values over 0.4 were removed in *Bos indicus* population. The following command can be used to generate the table with estimated Pi-hat value for pairwise individuals:

   ```
   plink --file pop --genome --out pop --cow --noweb.
   ```

2. The command "disjoin" in R package (GenomicRanges) will cut the CNVs according to the CNV boundaries. During the disjoin process, the first disjoined CNV segments are used to disjoin with other CNVs or CNV segments that will produce 1~2 bp small pieces. Those small artificial CNV segments are filtered by the command line: good_ CNV_segment <- raw_ CNV_segment[which(raw_CNV_segment[,3]- raw_CNV_ segment[,2]>=50),]. Here, we also filtered the CNVs segments smaller than 50 bp.
3. For more detailed information, one can refer to https://github.com/stephenturner/qqman.
4. Please note, if you used the R package to produce CNV segments, the boundaries of each segment should be recovered to the position of corresponding probe position before merging the CNV segments into CNVRs.
5. Numerous browsers can be used to detect genes overlapping or located in signature selection regions. Two of the widely used browsers are UCSC Genome Browser (http://genome.ucsc.edu/) and Ensembl (http://www.ensembl.org/index.html).
6. GeneCard (http://www.genecards.org/), NCBI (https://www.ncbi.nlm.nih.gov/), and many other databases can be used to track the single gene functions. For large groups of genes, one can use DAVID (https://david.ncifcrf.gov/) or other gene ontology tools to detect the potentially enriched gene clusters and functions.
7. The position information in indicus_BAF and taurine_BAF are the serial numbers of animals for each SNP and normalized by a scale according to the data.
8. In a diploid genome, the two alleles of each SNP are referred to as A and B alleles according to specific rules (https://www.

illumina.com/documents/products/technotes/technote_topbot.pdf). The LRR is referred as the signal intensity of a probe. PennCNV mainly integrates the BAF and LRR to identify CNV region and the copy number of the CNV. The LRR is normalized and will be around 0 in the normal region (two copies in a diploid genome). When the target region is a deletion, the probe cannot bind or will have less affinity to the target region. Correspondingly, the value of BAF of that SNP will be random. For more detail information, please refer to a PennCNV document (http://www.openbioinformatics.org/penncnv/advanced/penncnv_snp_allele.html) and the PennCNV paper (https://www.ncbi.nlm.nih.gov/pmc/articles/PMC2045149/).

9. The cattle UMD3.1 assembly covers 29 autosomes and the X chromosome and contains many gaps and sequences on unplaced scaffolds [13]. Although UMD3.1 has no chrY, it does contain sequence reads from both the cow (female, L1 Dominette 01449) and its sire (male, L1 Domino 99,375). Most of chrY sequences are found on unplaced scaffolds but occasionally some of them could be misplaced on autosomes or chrX. These misassignment errors would cause distorted dosage ratios when comparing between males/bulls and females/cows, thus leading to false positive CNVs.
10. The normal LRR of probes in the misassemblies from the Y chromosome to autosomes that showed no difference with the LRR of probes in the autosomes in bulls are normalized as "0". But the BAF will be around "0" or "1" ideally. In cows, the LRR value of probes in the misassemblies from the Y chromosome to autosomes should be negative since they do not contain Y chromosome. Correspondingly, the BAF values of these probes will be random.

References

1. Handsaker RE et al (2015) Large multiallelic copy number variations in humans. Nat Genet 47(3):296–303
2. Liu GE et al (2010) Analysis of copy number variations among diverse cattle breeds. Genome Res 20(5):693–703
3. Klambauer G et al (2012) Cn. MOPS: mixture of Poissons for discovering copy number variations in next-generation sequencing data with a low false discovery rate. Nucleic Acids Res 40(9):e69. https://doi.org/10.1093/nar/gks003
4. Li J et al (2012) CONTRA: copy number analysis for targeted resequencing. Bioinformatics 28(10):1307–1313
5. Kim J-H et al (2012) CNVRuler: a copy number variation-based case–control association analysis tool. Bioinformatics 28(13):1790–1792
6. Glessner JT, Li J, Hakonarson H (2013) ParseCNV integrative copy number variation association software with quality tracking. Nucleic Acids Res 41(5):e64. https://doi.org/10.1093/nar/gks1346
7. Xu L et al (2013) Comparative analysis of CNV calling algorithms: literature survey and a case study using bovine high-density SNP data. Microarrays 2(3):171–185
8. Zhou Y et al (2016) Comparative analyses across cattle genders and breeds reveal the pitfalls caused by false positive and lineage-

differential copy number variations. Sci Rep 6:29219

9. Zhou Y et al (2016) Genome-wide CNV analysis reveals variants associated with growth traits in Bos indicus. BMC Genomics 17(1):419
10. Park C et al (2017) A copy number variation in PKD1L2 is associated with colorectal cancer predisposition in korean population. Int J Cancer 140(1):86–94
11. Winchester L, Yau C, Ragoussis J (2009) Comparing CNV detection methods for SNP arrays. Brief Funct Genomic Proteomic 8(5):353–366. https://doi.org/10.1093/bfgp/elp017
12. Wang K et al (2007) PennCNV: an integrated hidden Markov model designed for high-resolution copy number variation detection in whole-genome SNP genotyping data. Genome Res 17(11):1665–1674
13. Elsik CG, Tellam RL, Worley KC (2009) The genome sequence of taurine cattle: a window to ruminant biology and evolution. Science 324(5926):522–528

Chapter 4

Whole-Genome Shotgun Sequence CNV Detection Using Read Depth

Fatma Kahveci and Can Alkan

Abstract

With the developments in high-throughput sequencing (HTS) technologies, researchers have gained a powerful tool to identify structural variants (SVs) in genomes with substantially less cost than before. SVs can be broadly classified into two main categories: balanced rearrangements and copy number variations (CNVs). Many algorithms have been developed to characterize CNVs using HTS data, with focus on different types and size range of variants using different read signatures. Read depth (RD) based tools are more common in characterizing large (>10 kb) CNVs since RD strategy does not rely on the fragment size and read length, which are limiting factors in read pair and split read analysis. Here we provide a guideline for a user friendly tool for detecting large segmental duplications and deletions that can also predict integer copy numbers for duplicated genes.

Key words Copy number variation, Whole genome shotgun sequencing, Read depth, mrFAST, mrsFAST

Abbreviations

CNV	Copy number variation
mrCaNaVaR	Micro read Copy Number Variant Regions
mrFAST	Micro read Fast Alignment Search Tool
mrsFAST	Micro read substitution only Fast Alignment Search Tool
RD	Read depth
TRF	Tandem repeat finder
WGS	Whole genome sequencing

1 Introduction

CNVs are changes in the amount of DNA in a genome that show themselves as duplication (gain) and deletion (loss) events [1]. Traditionally, a copy number variant is defined as a gain or loss in the amount of DNA in the genome that is larger than 1 Kb,

Derek M. Bickhart (ed.), *Copy Number Variants: Methods and Protocols*, Methods in Molecular Biology, vol. 1833, https://doi.org/10.1007/978-1-4939-8666-8_4, © Springer Science+Business Media, LLC, part of Springer Nature 2018

although smaller deletions or duplications (>50 bp) are considered CNV in several projects such as the 1000 Genomes Project. CNVs are known to have played major role in evolution [2–4], and several of them are correlated with complex human disease [5, 6].

There exist many algorithms to characterize CNVs in genomes that employ different strategies to characterize different classes of CNVs in different size ranges. Here we describe the mrCaNaVaR algorithm [7] that uses read depth information to predict large (>10 Kb) duplications and deletions, along with integer copy numbers for genes. mrCaNaVaR was the first tool developed solely for accurate assessment of segmental duplications on personal genomes, and it also was the only publicly available tool for integer copy number prediction. Although it was initially developed to analyze only human segmental duplications, it has been used to characterize large CNVs in many organisms including cattle [8], great apes [3, 9], Neandertal [10] and Denisovan [11, 12] ancient DNA, and even plant genomes [13].

Here we describe how to use mrCaNaVaR for detecting large CNVs. Since segmental duplication prediction necessitates tracking multiple map locations of reads, we need to use a multi-mapper [14]. Briefly, after mapping all reads to a repeat masked genome (to "clean up" common repeats) using mrFAST or mrsFAST [15, 16], mrCaNaVaR loads the mapping information (i.e., SAM files) and applies a sliding window strategy to count read depth. The read depth distribution is often affected by G + C content of genomic segments, therefore mrCaNaVaR applies a statistical smoothing method (LOESS) to correct for bias in high and low G + C regions. It then identifies those regions with higher or lower read depth than the genome-wide average (±3 standard deviations), and chains consecutive such regions. mrCaNaVaR also predicts integer copy numbers of windows of size 1 Kb as the ratio of observed read depth to average read depth, and uses this information to refine CNV predictions and the estimated breakpoints.

2 Materials

This section contains a list of prerequisite software tools that mrCaNaVaR requires. All tools described below are open source and free to use under the GNU public license.

2.1 Installing Zlib

1. Download the zlib program from https://sourceforge.net/projects/libpng/files/zlib/. An example of a Linux command to use is as follows (*see* **Notes 1**, **2**, **3**, and **4**):

   $ wget https://sourceforge.net/projects/libpng/files/zlib/X.X.X/zlib-X.X.X.tar.gz

   where X.X.X denotes version number.

2. Unpack the downloaded file using the command shown below:

   $ tar -xzvf zlib-X.X.X.tar.gz

   where X.X.X denotes version number.

   There should be a new folder named zlib-X.X.X in the current working directory.

3. Remove the zipped version of file:

   $ rm zlib-X.X.X.tar.gz

4. Change directory into zlib-X.X.X:

   $ cd zlib-X.X.X

5. Type at the shell prompt:

   $ ./configure

   $ make test

   If everything goes well, then type:

   $ sudo make install

Depending on the version of the GCC compiler and some additional library requirements, the installation should be completed without any errors (*see* **Note 5**).

2.2 Installing Blast

1. Download the Blast program from ftp://ftp.ncbi.nlm.nih.gov/blast/executables/LATEST/. An example of a Linux command to use is as follows:

   $ wget ftp://ftp.ncbi.nlm.nih.gov/blast/executables/LATEST/ncbi-blast-X.X.X+-x64-linux.tar.gz

   where X.X.X denotes version number.

2. Unpack the downloaded file using the command shown below:

   $ tar -xzvf ncbi-blast-X.X.X+-x64-linux.tar.gz

   where X.X.X denotes version number.

   There should be a new folder named ncbi-blast-X.X.X+ in the current working directory.

3. Remove the zipped version of file:

   $ rm ncbi-blast-X.X.X+-x64-linux.tar.gz

4. Change directory into ncbi-blast-X.X.X+/bin:

   $ cd ncbi-blast-X.X.X+/bin

5. Download ncbi-rmblastn from http://ftp.ncbi.nlm.nih.gov/blast/executables/rmblast/.

   $ wget ftp://ftp.ncbi.nlm.nih.gov/blast/executables/rmblast/X.X.X/ncbi-rmblastn-X.X.X-x64-linux.tar.gz

   where X.X.X denotes version number.

6. Unpack the downloaded file using the command shown below:

   $ tar -xzvf ncbi-rmblastn-X.X.X-x64-linux.tar.gz

where X.X.X denotes version number.

There should be a new folder named ncbi-rmblastn-X.X.X in the current working directory.

7. Copy rmblastn file into /path/ncbi-blast-X.X.X+/bin directory:

   $ cp /path/ncbi-rmblastn-X.X.X/rmblastn /path/ncbi-blast-X.X.X+/bin

8. Remove unused files in ncbi-blast-X.X.X+/bin directory:

   $ rm ncbi-rmblastn-X.X.X*

2.3 Installing Tandem Repeats Finder

1. Change directory into RepeatMasker:

   $ cd /path/RepeatMasker

2. Download "*trf*" file from http://tandem.bu.edu/trf/trf409.linux64.download.html.
3. Convert binary file into executable:

   $ chmod 755 trf409.linux64

4. Change file name:

   $ mv trf409.linux64 trf

2.4 Installing RepeatMasker

1. Download the RepeatMasker program from http://www.repeatmasker.org. An example of a Linux command to use is as follows:

   $ wget http://www.repeatmasker.org/RepeatMasker-open-X-X-X.tar.gz

   where X-X-X denotes version number.

2. Unpack the downloaded file using the command shown below:

   $ tar -xzvf RepeatMasker-open-X-X-X.tar.gz

   There should be a new folder named RepeatMasker in the current working directory.

3. Remove the zipped version of file:

   $ rm RepeatMasker-open-X-X-X.tar.gz

4. Change directory into RepeatMasker:

   $ cd RepeatMasker

5. RepeatMasker provides two open databases, Dfam and Dfam_consensus, and will work with these datasets, but it is advised to obtain a license for the RepBase RepeatMasker Edition to supplement these sequences. To obtain a license and download the library go to http://www.girinst.org. Copy RepBase into the RepeatMasker directory:

   $ cp RepBaseRepeatMaskerEdition-XXXXXXXX.tar.gz /path/RepeatMasker/

6. Change directory into:

   $ cd /path/RepeatMasker

7. Unpack the downloaded file using the command shown below:

   $ tar -xzvf RepBaseRepeatMaskerEdition-########.tar.gz

   There should be a new folder named Libraries in the current working directory.

8. Remove the zipped version of file:

   $ rm RepBaseRepeatMaskerEdition-########.tar.gz

9. Check for Dfam Updates (optional).

   (a) Change directory into Libraries:

   (b) Download the Dfam.hmm.gz file from http://www.dfam.org.

      $ wget http://www.dfam.org/web_download/Current_Release/Dfam.hmm.gz

   (c) Unpack the downloaded file using the command shown below:

      $ gunzip Dfam.hmm.gz

   There should be a new file named Dfam.hmm in the current working directory.

10. Change directory into RepeatMasker:

    $ cd /path/RepeatMasker/

11. Type at the shell prompt (*see* **Note 6**):

    $ perl ./configure

12. Follow the instructions.

2.5 Installing mrFAST

1. Download the mrFAST program from https://github.com/BilkentCompGen/mrfast/. An example of a Linux command to use is as follows (*see* **Notes 1**, **2**, **3**, and **4**):

   $ wget https://github.com/BilkentCompGen/mrfast/archive/vX.X.X.X.zip

   where X.X.X.X denotes version number.

2. Unpack the downloaded file using the command shown below:

   $ tar -xzvf mrfast-X.X.X.X.tar.gz

   where X.X.X.X denotes version number. There should be a new folder named mrfast-X.X.X.X in the current working directory.

3. Remove the downloaded compressed version of the file:

   $ rm mrfast-X.X.X.X.tar.gz

   where X.X.X.X denotes version number.

4. Change directory into mrfast-X.X.X.X:

$ cd mrfast-X.X.X.X

where X.X.X.X denotes version number.

5. Type at the shell prompt:

   $ make

6. Make it executable from anywhere in terminal using the commands shown below (optional) or you will give the path of mrFAST tool to run:

   $ nano ~/.bashrc

   Add the current directory of mrFAST at the end of the file using the following command:

   export PATH=$PATH:/mrFAST_directory/

   Save the file and close. Then, run the following command to make it permanent:

   $ source ~/.bashrc

2.6 Installing mrsFAST (Alternative to mrFAST Above)

1. Install git:

   $ sudo apt install git

   this assumes that the underlying Linux OS is Debian-based. For other Linux distributions, refer to official git page at https://git-scm.com/.

2. Download the mrsFAST program from https://github.com/sfu-compbio/mrsfast. An example of a Linux command to use is (*see* **Notes 1**, **2**, **3**, and **4**):

   $ git clone https://github.com/sfu-compbio/mrsfast

3. Change directory into mrsfast:

   $ cd mrsfast

4. Type at the shell prompt:

   $ make

5. Make it executable from anywhere in terminal using the commands shown below (optional) or you will give the path of mrsFAST tool to run:

   $ nano ~/.bashrc

   Add the current directory of mrsFAST at the end of the file using the following command:

   export PATH=$PATH:/mrsFAST_directory/

   Save the file and close. Then, run the following command to make it permanent:

   $ source ~/.bashrc

2.7 Installing mrCaNaVaR

1. Download the mrCaNaVaR program from https://github.com/BilkentCompGen/mrcanavar. An example command line is (*see* **Notes 1** and 7):

   $ git clone https://github.com/BilkentCompGen/mrcanavar

2. Change the working directory to mrcanavar

   $ cd mrcanavar

3. Compile by typing make:

   $ make

4. Make it executable from anywhere in terminal using the commands shown below (optional) or you will give the path of mrCaNaVaR tool to run:

   $ nano ~/.bashrc

   Add the current directory of mrCaNaVaR at the end of the file using the following command:

   export PATH=$PATH:/mrCaNaVaR_directory/

   Save the file and close. Then, run the following command to make it permanent:

   $ source ~/.bashrc

3 Methods

In the first preprocessing step, common and tandem repeats will be masked with N characters in reference genome assembly using RepeatMasker [17] and Tandem Repeats Finder [18] tools. Here lowercase letters will be treated as nonrepeats. For the human genome assemblies, we recommend to use both:

- *RepeatMasker (with sensitive "-s" parameter enabled):* The repeat masked versions of many genome assemblies are available at the UCSC Genome Browser.
- *Tandem Repeats Finder* (parameters: trf input_file 2 7 7 80 10 50 500 -m)

In the second preprocessing step, mrFAST (or mrsFAST) search index will be built using the repeat masked version of the same reference genome (*see* **Note 8**). For mrFAST, you can use the default 12 character-window size without losing the sensitivity if your calculated window size is greater than default.

$ ./mrfast --index $(reference_genome).fasta , OR

$ ./mrsfast --index $(reference_genome).fasta

Index file will be created in the same directory upon the completion of the indexing phase. The next step using samtools is optional. If skipped, merely the output SAM files will not include headers. You can create $(reference_genome).masked.fai using samtools:

$ samtools faidx $(reference_genome).masked.fa

In the third preprocessing step, reads will be mapped to the reference genome using mrFAST (or mrsFAST) in single-read mapping mode, and *not* in paired-end mapping mode. Here mrFAST (or mrsFAST) may be used for read mapping. Although other aligners might also be used if all map locations, including suboptimal alignments, are reported, no other mapper is recommended to use with mrCaNaVaR for characterization of duplications and absolute copy number prediction. It is strongly recommended to use mrFAST (or mrsFAST) that was specifically developed for SV/CNV analysis. The same repeat masked version of the reference genome described above will be used for building the mrFAST (or mrsFAST) index and read mapping. FASTA and index file of the reference genome should be found in the same folder. mrFAST reports all the locations per read by default. Here recommended edit/hamming distance threshold is 5% of the read length (i.e., 2 for 36 bp reads).

$ mrfast --search $(reference_genome).fasta --seq $(input_file).fastq -e $(distance_threshold).

The reported locations will be saved into "output" by default. If you want to save it somewhere else, you can use "-o" parameter to specify another file.

$ mrfast --search $(reference_genome).fasta --seq $(input_file).fastq -o $(output_directory).

Both mrFAST and mrsFAST are capable of generating map files in plain and gzip-compressed SAM format.

mrCaNaVaR has four running modes.

3.1 Preparing Reference--prep

Users need to run this mode only once for a new reference genome assembly unless repeat masking or the sliding window sizes are altered. In this mode, coordinates and G + C content of three sliding window classes (Long Windows (LWs), Short Windows (SWs), and Copy Windows (CWs)) over the reference genome are calculated and saved in a *genome configuration* file ($(reference_genome).cnvr) (Table 1).

Sliding windows: mrCaNaVaR operates on three classes of sliding windows over the reference genome to estimate absolute copy numbers and discover duplicated and deleted regions. The three windows are the following:

1. *Long Windows (LWs)* are used to find locations of deletions and duplications.

The LWs are overlapping windows of size 5 kbp of unmasked characters (i.e. A, C, G, T only), sliding by 1 kbp of any characters. For example, in repeat masked GRCh37 assembly, the first three LWs are:

chr1	10,000	17,048
chr1	11,000	17,048
chr1	12,000	17,236

Table 1
mrCaNaVaR in --prep mode

Sample command line
$ mrcanavar --prep -fasta $(reference_genome)_masked.fasta -gaps $(reference_genome)_gaps.bed -conf $(reference_genome).cnvr
Inputs
--fasta: Repeat masked reference genome (FASTA format)
-gaps: Gap coordinates of the reference genome (BED format)
Outputs
-conf: Reference configuration file (a native binary format)
Optional parameters
-lw_size <lw_size>: Long window span size. Default is 5000 bp of non-masked characters which are A, C, G, and T.
-lw_slide <lw_slide>: Long window slide size. Default is 1000.
-sw_size <sw_size>: Short window span size. Default is 1000.
-sw_slide <sw_slide>: Short window slide size. Default is 1000.
-cw_size <cw_size>: Copy number window size. Default is 1000.
-pseudoa <bed_file>: Coordinates for pseudoautosomal regions in the reference genome (BED format).

In the example above, the first window has 7048 any characters, but 5000 non-N characters (thus 2048 N characters). We then slide the window by 1000 *any* base pairs, starting the second window at position 11,000. The end of the second window is again 17,048 since all of the first 1000 characters of chr1 are masked. The third window starts at position 12,000 and contains 5326 any characters, where 5000 are non-N.

2. *Short Windows (SWs)* are used to refine breakpoints of detected duplication and deletion intervals. Default window size is 1000 bp of nonmasked characters, and the slide size is 1000 bp of any characters, similar to LW.
3. *Copy Windows (CW)* are nonoverlapping windows and used to predict absolute copy numbers. Default window size is 1000 bp of nonmasked characters.

3.2 Reading/Loading Alignments--read

In this mode, mapping information in SAM format which is either uncompressed (*.sam) or compressed with gzip (*.sam.gz) is loaded, the read depth in LW, SW, and CW intervals are calculated, and read depth information is be saved in a *depth* output file in a native binary format (Table 2). Additionally, three human-readable text files are generated for the raw and G + C corrected read depths and G + C contents of LW, SW, and CW intervals. These three

Table 2
mrCaNaVaR in --read mode

Sample command line
$ mrcanavar --read -conf $(reference_genome).cnvr --gz -samdir $(sam_file_directory) -depth $(sample).depth
$ mrcanavar --read -conf $(reference_genome).cnvr --gz -samdir $(sam_files_directory) -depth $(sample).mult.depth --multgc
Inputs
-conf: Reference configuration file created using the PREP mode.
-samdir: Directory of plain SAM files (*.sam) or gzip-compressed SAM files (*.sam.gz)
Outputs
- **$(sample).depth:** Read depth file in a native binary format.
Optional parameters
--gz: Indicates the SAM files are compressed in gzip format. **-nsam <first> <second>:** Choose a sequence of SAM files between two indexes.

Table 3
mrCaNaVaR in --conc mode

Sample command line
$ mrcanavar --conc -conf $(reference_genome).cnvr -concdepth lib1.depth lib2.depth lib3.depth -depth merged.depth
Inputs
-conf: Reference configuration file created using the PREP mode.
-concdepth: Read depth file(s) to merge.
Outputs
-depth: Read depth file in a native binary format.

human readable text files are generated for manual inspection and debugging purposes only, if needed, and will not be loaded and needed by mrCaNaVaR when calling CNVs.

3.3 Merging Multiple Libraries--conc if they Exist

The CONC mode is optional and needed only if the SAM files are read into a number of *.depth files or if there are more than one library preparation which should be read into different *.depth files. In this mode, mrCaNaVaR will be used to merge more than 2 *.depth files using the READ mode and output the merged LW, SW, and CW intervals (Table 3).

3.4 Calling CNVs and Copy Numbers--call

In this mode, segmental duplication and deletion intervals are called and absolute copy numbers are predicted. Absolute copy number over the CW intervals, and duplicated and deleted regions in the analyzed sample using the LW and SW intervals will be predicted. Additionally, if a BED file that contains genes in the reference genome is provided, the median and average copy numbers of the intervals spanning the genes are generated. We recommend using the median copy number values for genes (Table 4).

Table 4
mrCaNaVaR in --call mode

Sample command line
$ mrcanavar --call -conf $(reference_genome).cnvr -depth dep $sample.depth -o $sample
Inputs
-conf: Reference configuration file created using the PREP mode.
-depth: Read depth file created using the READ mode.
-o: Prefix for the output file names.
Outputs
- **$(sample).log:** Log file that describes what mrCaNaVaR did including read depth statistics.
- **$(sample).copynumber.bed:** Estimated copy numbers over CW intervals.
- **$(sample).dels.bed:** Estimated deleted regions over CW intervals.
- **$(sample).dups.bed:** Estimated duplicated regions over CW intervals.
- **$(sample).cw_norm.bed:** GC-corrected read depth values for the inferred control regions for CW intervals.
- **$(sample).lw_norm.bed:** GC-corrected read depth values for the inferred control regions for LW intervals.
- **$(sample).sw_norm.bed:** GC-corrected read depth values for the inferred control regions for SW intervals.
Optional Parameters
--xx: Set gender of the sequenced sample as female. Mammalian genomes only. Default is autodetect.
--xy: Set gender of the sequenced sample as male. Mammalian genomes only. Default is autodetect.
--multgc: Perform multiplicative GC correction. Default is additive.
-mindup <min_dup_len>: Minimum duplication length. Default is 10000.
-gene <genelist>: Coordinates for genes calculating gene-based copy numbers.
--verbose: Verbose output.

4 Notes

1. wget is a computer program that retrieves content from web servers.
2. Perl version 5.8.0 or higher installed is required.
3. mrFAST with FastHASH (post 2.5.0.0) is drastically faster than plain mrFAST (pre 2.5.0.0). When using mrFAST, please ensure that the version is >2.5.0.0.
4. zlib is for the ability to read compressed SAM files.
5. Installing programs in Unix like systems requires admin privileges.
6. The program requires some initial configuration. This should also be rerun after future updates to the library files.
7. mrCaNaVaR is developed with gcc version 4.1.2 (C compiler)
8. mrsFAST is multithreaded and faster than mrFAST. Running mrsFAST is similar to mrFAST. Manual for mrsFAST is found at https://github.com/sfu-compbio/mrsfast//blob/master/README.md.

References

1. Alkan C, Coe BP, Eichler EE (2011) Genome structural variation discovery and genotyping. Nat Rev Genet 12:363–376
2. Ventura M, Catacchio CR, Alkan C et al (2011) Gorilla genome structural variation reveals evolutionary parallelisms with chimpanzee. Genome Res 21:1640–1649
3. Prado-Martinez J, Sudmant PH, Kidd JM et al (2013) Great ape genetic diversity and population history. Nature 499:471–475
4. Sudmant PH, Huddleston J, Catacchio CR et al (2013) Evolution and diversity of copy number variation in the great ape lineage. Genome Res 23:1373–1382
5. Sharp AJ, Cheng Z, Eichler EE (2006) Structural variation of the human genome. Annu Rev Genomics Hum Genet 7:407–442
6. Sharp AJ, Hansen S, Selzer RR et al (2006) Discovery of previously unidentified genomic disorders from the duplication architecture of the human genome. Nat Genet 38:1038–1042
7. Alkan C, Kidd JM, Marques-Bonet T et al (2009) Personalized copy number and segmental duplication maps using next-generation sequencing. Nat Genet 41:1061–1067
8. Bickhart DM, Hou Y, Schroeder SG et al (2012) Copy number variation of individual cattle genomes using next-generation sequencing. Genome Res 22:778–790
9. Prado-Martinez J, Hernando-Herraez I, Lorente-Galdos B et al (2013) The genome sequencing of an albino Western lowland gorilla reveals inbreeding in the wild. BMC Genomics 14:363
10. Green RE, Krause J, Briggs AW et al (2010) A draft sequence of the Neandertal genome. Science 328:710–722
11. Reich D, Green RE, Kircher M et al (2010) Genetic history of an archaic hominin group from Denisova Cave in Siberia. Nature 468:1053–1060
12. Meyer M, Kircher M, Gansauge M-T et al (2012) A high-coverage genome sequence from an archaic Denisovan individual. Science 338:222–226
13. Cardone MF, D'Addabbo P, Alkan C et al (2016) Inter-varietal structural variation in grapevine genomes. Plant J 88:648–661
14. Chiang DY, McCarroll SA (2009) Mapping duplicated sequences. Nat Biotechnol 27:1001–1002
15. Hach F, Hormozdiari F, Alkan C et al (2010) mrsFAST: a cache-oblivious algorithm for short-read mapping. Nat Methods 7:576–577
16. Hach F, Sarrafi I, Hormozdiari F et al (2014) mrsFAST-ultra: a compact, SNP-aware mapper for high performance sequencing applications. Nucleic Acids Res 42:W494–W500
17. Smit AFA, Hubley R, Green P (1996 2004) RepeatMasker Open-3.0
18. Benson G (1999) Tandem repeats finder: a program to analyze DNA sequences. Nucleic Acids Res 27:573–580

Chapter 5

Read Depth Analysis to Identify CNV in Bacteria Using CNOGpro

Ola Brynildsrud

Abstract

Whole-genome sequencing with short-read technologies is well suited for calling single nucleotide polymorphisms, but has major problems with the detection of structural variants larger than the read length. One such type of variation is copy number variation (CNV), which entails deletion or duplication of genomic regions, and the expansion or contraction of repeated elements. Duplicated and deleted regions will typically be collapsed during de novo assembly of sequence data, or ignored when mapping reads toward a reference. However, signatures of the copy number variation can be detected in the resultant read depth at each position in the genome. We here provide instructions on how to analyze this read depth signal with the R package CNOGpro, allowing for estimation of copy numbers with uncertainty for each feature in a genome.

Key words Read depth, CNV, Bacteria, CNOGpro, Coverage, Whole-genome sequencing

1 Introduction

Copy number variation (CNV) is gaining recognition as an important contributor to bacterial phenotypes [1, 2]. CNVs are variations in the number of times a particular sequence element appears from one genome to another. Technically, insertion/deletion events (indels), microsatellites, and even mononucleotide repeats are types of CNV, but the term CNV is normally applied to longer features such as genes. Although there is no accepted standard for how long a feature needs to be in order to be called a CNV, there is an interesting methodological schism between features that are shorter and features that are longer than a sequence read. Due to the highly accurate nature of short-read technologies like Illumina, CNVs that are shorter than a single sequence read are trivial to resolve, since they will be represented in full within the read. Similarly, CNVs that fall in their entirety within the boundaries of paired reads are relatively simple to resolve, and de novo assemblers that consider insert size should be able to reconstruct the

Derek M. Bickhart (ed.), *Copy Number Variants: Methods and Protocols*, Methods in Molecular Biology, vol. 1833, https://doi.org/10.1007/978-1-4939-8666-8_5,

region with full accuracy. However, CNVs that are significantly longer than this are unresolvable in standard de novo assembly. Similarly, when mapping reads against a reference genome, CNV differences between the sample and the reference are usually ignored unless the feature is contained within a single read. However, in both these approaches, a signature of CNV remains in the read depth at each position. This signal can be analyzed in order to determine the copy number of each sequence region.

The following assumptions are made: (1). In short read sequencing, the number of reads generated at each position of a sequenced genome is approximately equal. (2) All reads can be perfectly and unambiguously mapped to a reference genome (which can be a de novo assembly of the same strain) that is identical to the strain being tested for CNV. (3) The read depth along each genomic feature in this mapping is proportional to the copy number of the feature in the sequenced genome. Only the last of these assumptions, (3), is generally fulfilled [3–5]. Assumption (1) is violated by a number of factors that influence read generation, like local GC content [6], probe GC content [7], replication bias/population heterogeneity [8], and lab/batch effects [9]. Assumption (2) is never fulfilled, and in practice read alignment is strongly affected by genomic mappability, a concept related to the uniqueness of a sequence feature [10]. Furthermore, mapping is affected by sequencing errors, structural rearrangements, and indels [11].

The following will present CNOGpro, an R package for quantifying copy numbers of sequence features in bacterial genomes [12]. This document will only present a typical workflow, with only very superficial description of the methods. For a deeper understanding of the tool I therefore refer to the 2015 paper [12].

The CNOGpro procedure starts with two files: A reference file in the GenBank format, and an alignment of reads toward this (or a FASTA representation of this file) sequence in the sorted SAM/BAM format. The latter file contains the genomic coordinates of all reads mapped to the reference. This information is read in a sliding window across the reference, with each window getting a read count. These read counts follow a negative binomial distribution with parameters depending on the copy number of the sequence feature. Counts are first corrected for GC bias (the GC enrichment in each window is calculated from the reference sequence), then put into either of two independent statistical models for estimating CNVs across the genome: (A) A hidden Markov model (HMM) that estimates copy numbers and CNV breakpoints linearly across the genome, but which is blind to the coordinates of coding segments across the genome. There is no confidence interval associated with the inferred number. This method would be superior for detecting CNV not restricted to any single sequence feature, e.g., a phage duplication or expansion of a tandem repeat sequence. (B) A function that pools consecutive read counts that are part of the same coding segment or intergenic region, and then uses the relative mean of these observations to the global mean of single-copy regions to estimate a copy number for the feature. A confidence interval around the CNV estimate is found by repeated sampling with replacement (bootstrapping). This method would be preferred for detecting duplicated or lost gene copies. The results from (A) and (B) can be interpreted separately or in unison, CNOGpro provides no way to distill the two into a single number.

2 Materials

CNOGpro is provided as an R package (R citation here) available through the CRAN repository (cran.r-project.org). To install it, run the following command in R (The greater than symbol - ">" - is just the prompt to show that your system is ready to accept commands, and should not be entered manually):

```
> install.packages("CNOGpro")
```

This should also automatically install the only dependency, the R package SeqInR [13]. In the following, I will show one possible way to prepare the input files.

2.1 Preparing a Reference File in the GenBank Format

There are basically two ways of doing this. The easiest is to search for an existing reference genome in the GenBank format. A good place to start would be NCBI's databases, like https://www.ncbi.nlm.nih.gov/genbank/. The closer related your reference genome is to your genome to be tested for CNV, the better. If there are no good references available, you might have to create your own. This can be done by running a gene predictor such as Prokka [14] on a FASTA format file you have. One option would be to first do a de novo assembly of your test genome, then feeding this assembly file (presumably in several contigs) to Prokka. An important restriction of CNOGpro is that it only accepts GenBank files with a single chromosome or contig. Therefore, if you have multiple contigs in a GenBank file, you would have to split it up. This can easily be done in a UNIX environment:

```
$ csplit --quiet --elide-empty-files CONTIGFILE
"/^\/\//+1" "{*}"
```

This will work with up to 100 contigs. If you have more than that, add --digits=3 to the command. One caveat is important: CNOGpro assumes a copy number of 1 in at least some parts of a chromosome/contig. Therefore, it is not straightforward to use CNOGpro on different contigs to estimate relative copy numbers, for example to find the copy number of a plasmid if the plasmid is on a separate contig.

2.2 Preparing a Sequence Alignment File and the Read Coordinates

Start by mapping your quality controlled reads against your reference (in FASTA format). Then sort your SAM file and extract read coordinates. A basic approach is demonstrated below. You will need the programs BWA [15], and samtools [16] installed, as well as the programming language perl:

```
$ bwa index CONTIGFILE.FASTA
$ bwa mem CONTIGFILE.FASTA FORWARDREADS.FASTQ
REVERSEREADS.FASTQ > aln.sam
$ samtools view -bS aln.sam > aln.bam && samtools sort
aln.bam > aln.sorted.bam
$ samtools view aln.sorted.bam | perl -lane 'print
\"$F[2]\t$F[3]\"' > aln.hits
```

Some users have had problems running the last command. This was solved by removing the backlashes from the print command. At this point you have everything you need to run a CNV analysis with CNOGpro. Start R and when the prompt is ready enter the following to load the CNOGpro package:

```
> library("CNOGpro")
```

3 Methods

CNOGpro has a sequential workflow in the R scripting language. You start by creating an R object that holds the necessary data, such as the genome sequence, and the name and coordinates of aligned reads. An additional GC-bias-corrected count can be made at this point. If the correction is performed, the bias-corrected counts will be used in all further analyses. CNV analysis proceeds with one or both of the HMM and the bootstrap method. Finally, there are built-in methods for storing the results to file, and some nifty plot methods that can provide a better understanding of the CNV landscape.

3.1 Creating the CNOGpro Object

All CNOGpro methods operate on a central object, unimaginatively called a CNOGpro object. In addition to requiring the (path to the) GenBank reference and read coordinate files, the user can also name their experiment strain and provide the size of the sliding windows used for counting read depth. The former has little impact beyond cosmetically in plots and other output, whereas the latter can make a huge difference. As a rule of thumb, if decent average read depth (30–100×) has been achieved, window lengths can be shorter than if the experiment resulted in poor read depth (<10×). Shorter window lengths increases sensitivity (to find true CNV regions) at a marginal cost to specificity (increased chance for detecting CNV where there really is no difference between the reference and experiment strain).

```
> experiment1 <- CNOGpro(hitsfile = "path/to/aln.hits",
gbkfile = "path/to/reference.gbk", windowlength = 100,
name = "Renibacterium salmoninarum str. Carson5b")
```

For now, the CNOGpro object "experiment1" is basically a complex type of list that holds a table of all the sequence features, the read depth in each 100 bp size window along the chromosome, and some other meta-information taken from the GenBank file.

3.2 Correcting for GC Bias

Before any statistical inferences are attempted, it is extremely important to correct for a bias in read depth introduced due to local GC content. GC content is thought to influence the efficiency of the PCR amplification in the library preparation and sequencing steps. The bias is unimodal, with both low-GC and high-GC fragments tending to have lower read counts than expected from a uniform coverage model [17]. The goal of GC bias normalization is thus to ensure that the median read depth is the same in low-GC and high-GC regions of the genome. In CNOGpro, we do this by inflating or deflating the read counts in regions of the genome with a GC skew [18]. The original observed counts are not overwritten and can still be used, but by default, all subsequent analyses will use GC-corrected counts. Correcting for this bias is a fully optional step, but the procedure is quick, and tends to improve the accuracy of CNV predictions. One important shortcoming is integrated phages, which tend to have a different GC content from the host genome and possibly also a variable copy number. However, unless the phage GC content is extremely different from the GC content of the host genome, it is still best to correct for GC bias in the read depth regions. Performing the command is as simple as feeding the CNOGpro object, as no extra parameters need to be set:

```
> experiment1.gcnorm <- normalizeGC(experiment1)
```

There are now multiple ways to proceed with CNV calling. The HMM and the bootstrap approach are completely independent of each other. You can run one of them, then store results, or you can run both.

3.3 Running the Hidden Markov Model (HMM)

Now that we have read counts and the important bits of a reference genome (coding segment breakpoints, genome length and GC content) loaded, CNV analysis can start. CNOGpro has implemented a HMM approach that allows copy number determination linearly along the genome. This procedure is blind to coding segment breakpoints, and determines copy number and breakpoint coordinates (i.e., the places in the genome where the copy number changes) from the read counts alone. As a result, it is better suited to discovering CNV within parts of a gene, in intergenic segments, in the expansion or contraction of linearly repeated elements, and across multigene regions such as integrated phages. However, their result is in discrete numbers (0, 1, 2, 3, etc.) and reported without uncertainty. The HMM approach proceeds in a sliding window fashion. For each window, the probability that the copy number of that window is 0, 1, 2, 3, etc. is calculated. This probability depends on the read depth and the copy number of the previous window. To set this up, an estimate must be made of the *transmission* probability (called *changeprob*), i.e., the probability of moving from one copy number to another. By default, this is 1.0E-4 for any transition, but this parameter is changeable. Determining the value of this parameter is unfortunately not trivial, and ideally a user has access to training data where the true copy numbers are known. As a rule of thumb though, higher transmission probability invokes more frequent change of the copy number state along the HMM chain. Another parameter that need to be set is *nstates*, which is the number of allowed copy number states. For example, setting nstates=5 would allow a copy number of 1, 2, 3, 4, and 5, but not higher. Similarly, setting *includeZeroState*=TRUE will add 0 to the list of allowed copy numbers. (This is highly recommended if there is any chance for a deletion in your sample.) If the copy number is zero, reads counts are not distributed as a negative binomial, but rather a geometric distribution that depends on the fraction of erroneously mapped reads, *errorRate*. By default, this is set to 1 out of every 1000 reads. Increasing this number means that more deletions will be called in regions with low (or no) coverage. With default parameters the command looks like this:

```
> experiment1.hmm <- runHMM(experiment1.gcnorm,
nstates=5, changeprob=0.0001,
includeZeroState = TRUE, errorRate=0.001)
```

This will add to the CNOGpro object a table with the most probable inferred copy number along the entire genome and the genomic coordinates where the copy number switches between states. It might say for example, that from coordinates 1–10,000 the copy number was 1, from 10,001 to 12,000 it was 2, and from 12,001 to 13,000 it was 0.

3.4 Running the Bootstrap Model for Gene-Wise Copy Number Estimates

A fundamentally different way to parse the read counts is to group them together by genome features such as coding sequences and intergenic regions. The read counts within such a feature follow an overdispersed count (Poisson) distribution [19], and the ratio of the mean of this distribution to the mean of a single-copy gene region gives the copy number. If the read depth within a single-copy gene averages to 50, we expect the read depth to average 100 in a duplicated gene, 150 in a triplicated gene, etc. However, the variance also grows with increasing copy number, so high-copy genes can have a very wide read count distribution.

CNOGpro uses the mean count within each sequence to make a point estimate of the copy number, and uses repeated sampling with replacement (bootstrapping) of the counts within the feature to estimate the associated uncertainty (the confidence interval) around the estimate. All parameters concern this bootstrapping procedure: The *replicates* parameter dictates the number of bootstrapped datasets per feature, and the *quantiles* parameter tells CNOGpro which quantiles of the distribution of bootstrapped estimates should be included in the results. The default is to include the 2.5 and 97.5-percentile of this distribution, i.e., a 95% confidence interval.

```
> experiment1.boot <- runBootstrap(experiment1.hmm,
replicates=1000, quantiles=c(0.025,0.0975))
```

Owing to the massive resampling, this method is relatively slow. A progress bar is provided to estimate remaining time.

Note: In a GenBank file, some sequence features can have a very complex architecture. For example, a single coding segment does not actually need to be contiguous, and can have additional, smaller coding segments nested within itself. This can lead to counterintuitive results for these genes. I advise that the bootstrap results are ignored for these pseudogenes.

3.5 Summarizing, Visualizing, and Storing Results

In the following I will describe methods for generating meaningful results from your CNOGpro experiment. First, there is the plotCNOGpro method, which creates several nifty figures, depending on what methods you have already ran on you CNOGpro object.

```
> plotCNOGpro(experiment1.boot)
```

If you have followed the commands in this chapter, the following output plots are shown (Press ENTER to see the next plot in the series): (1) a scatter plot of (corrected) read counts along the chromosome/contig. (2) The distribution (kernel density plot) of read counts in each copy number state, as assigned by the HMM method. (3) Box plots of the read counts within each GC content percentile, before and after normalization.

The printCNOGpro method will output either a summary of your HMM analysis, or the full table from your bootstrap analysis if you have not performed HMM analysis. If you have performed neither it will give you an error message.

```
> printCNOGpro(experiment1.boot)
```

Finally, the store method will store all your results to a file. Here you can set outputEntireTable to FALSE in order to just print the HMM results. The path parameter is used to set the path of the output file. The file is named according to the name parameter given to it when the experiment was initiated with the

CNOGpro command, in this case "Renibacterium salmoninarum str. Carson5b.txt":

```
> store(experiment1.boot, outputEntireTable=TRUE,
path="/path/to/output/")
```

The output file is a tab-separated table with all relevant results from the experiment. The columns are explained in Table 1.

Table 1
Explanation of CNOGpro output columns and possible values

Column	Description
Type	The type of sequence feature, as taken from the GenBank file. CDS = coding segment. IG = intergenic. tRNA and rRNA are special categories
Locus	The locus tag, as read from the GenBank file
Strand	Whether the orientation of the feature is in the forward (1) or reverse (−1) direction. Intergenic regions are always counted as forward
Left	The leftmost coordinate of the sequence feature, i.e., start if in forward direction, stop if in reverse.
Right	The rightmost coordinate of the sequence feature, i.e., stop if in forward direction, start if in reverse.
Length	Length of the sequence feature
CN_HMM	Copy number according to the HMM. Multiple values possible if breakpoints were located within the feature.
CN_boot	Floating point copy number estimate from the runBootstrap method
Lowerbound	The lower bound of the confidence interval around CN_boot. (By default, the 2.5 percentile in the distribution of bootstrapped estimates)
Upperbound	The upper bound of the confidence interval around CN_boot. (By default, the 97.5 percentile in the distribution of bootstrapped estimates)

References

1. Riehle MM, Bennett AF, Long AD (2001) Genetic architecture of thermal adaptation in Escherichia coli. Proc Natl Acad Sci 98(2):525–530
2. San Millan A, Escudero JA, Gifford DR et al (2016) Multicopy plasmids potentiate the evolution of antibiotic resistance in bacteria. Nat Ecol Evol 1:0010
3. Alkan C, Kidd JM, Marques-Bonet T et al (2009) Personalized copy number and segmental duplication maps using next-generation sequencing. Nat Genet 41(10):1061–1067
4. Campbell PJ, Stephens PJ, Pleasance ED et al (2008) Identification of somatically acquired rearrangements in cancer using genome-wide massively parallel paired-end sequencing. Nat Genet 40(6):722–729
5. Medvedev P, Stanciu M, Brudno M (2009) Computational methods for discovering structural variation with next-generation sequencing. Nat Methods 6:S13–S20
6. Dohm JC, Lottaz C, Borodina T et al (2008) Substantial biases in ultra-short read data sets from high-throughput DNA sequencing. Nucleic Acids Res 36(16):e105-e105
7. Diskin SJ, Li M, Hou C et al (2008) Adjustment of genomic waves in signal intensities from whole-genome SNP genotyping platforms. Nucleic Acids Res 36(19):e126-e126
8. Skovgaard O, Bak M, Løbner-Olesen A et al (2011) Genome-wide detection of chromosomal rearrangements, indels, and mutations in circular chromosomes by short read sequencing. Genome Res 21(8):1388–1393
9. Khrameeva EE, Gelfand MS (2012) Biases in read coverage demonstrated by interlaboratory and interplatform comparison of 117 mRNA and genome sequencing experiments. BMC Bioinformatics 13(6):S4
10. Lee H, Schatz MC (2012) Genomic dark matter: the reliability of short read mapping illustrated by the genome mappability score. Bioinformatics 28(16):2097–2105
11. Sims D, Sudbery I, Ilott NE et al (2014) Sequencing depth and coverage: key considerations in genomic analyses. Nat Rev Genet 15(2):121–132
12. Brynildsrud O, Snipen L-G, Bohlin J (2015) CNOGpro: detection and quantification of CNVs in prokaryotic whole-genome sequencing data. Bioinformatics 31(11):1708–1715
13. Charif D, Lobry JR (2007) SeqinR 1.0-2: a contributed package to the R project for statistical computing devoted to biological sequences retrieval and analysis. In: Structural approaches to sequence evolution. Springer, New York, pp 207–232
14. Seemann T (2014) Prokka: rapid prokaryotic genome annotation. Bioinformatics 30(14):2068–2069
15. Li H (2013) Aligning sequence reads, clone sequences and assembly contigs with BWA-MEM. *arXiv:1303.3997* [q-bio.GN]
16. Li H, Handsaker B, Wysoker A et al (2009) The sequence alignment/map format and SAMtools. Bioinformatics 25(16):2078–2079
17. Benjamini Y, Speed TP (2012) Summarizing and correcting the GC content bias in high-throughput sequencing. Nucleic Acids Res 40(10):e72-e72
18. Yoon S, Xuan Z, Makarov V et al (2009) Sensitive and accurate detection of copy number variants using read depth of coverage. Genome Res 19(9):1586–1592
19. Sepúlveda N, Campino SG, Assefa SA et al (2013) A Poisson hierarchical modelling approach to detecting copy number variation in sequence coverage data. BMC Genomics 14(1):128

Chapter 6

Using HaMMLET for Bayesian Segmentation of WGS Read-Depth Data

John Wiedenhoeft and Alexander Schliep

Abstract

CNV detection requires a high-quality segmentation of genomic data. In many WGS experiments, sample and control are sequenced together in a multiplexed fashion using DNA barcoding for economic reasons. Using the differential read depth of these two conditions cancels out systematic additive errors. Due to this detrending, the resulting data is appropriate for inference using a hidden Markov model (HMM), arguably one of the principal models for labeled segmentation. However, while the usual frequentist approaches such as Baum-Welch are problematic for several reasons, they are often preferred to Bayesian HMM inference, which normally requires prohibitively long running times and exceeds a typical user's computational resources on a genome scale data. HaMMLET solves this problem using a dynamic wavelet compression scheme, which makes Bayesian segmentation of WGS data feasible on standard consumer hardware.

Key words HaMMLET, Hidden Markov Model, Bayesian inference, CNV, Whole genome sequencing, Segmentation

1 Introduction

HaMMLET is a fast and memory-efficient method for approximate Bayesian inference of the posterior state marginals in a hidden Markov model (HMM) on large-scale data [1]. It allows for highly accurate inference of copy-number variance (CNV), using very limited computational resources.

Due to their many advantages over frequentist techniques [2–4], Bayesian HMM methods [5, 6] have repeatedly been proposed for CNV inference (e.g., [7–9]). In general, however, such MCMC approaches are very slow, memory-intensive, and prohibitive for genome-sized data. HaMMLET achieves large improvements in convergence, speed and memory by compressing the data into blocks using techniques based on Haar wavelet regression. Depending on the parameter settings, segmentation of WGS read-depth data can be performed within a few minutes on consumer hardware.

Derek M. Bickhart (ed.), *Copy Number Variants: Methods and Protocols*, Methods in Molecular Biology, vol. 1833,
https://doi.org/10.1007/978-1-4939-8666-8_6,

Given some prior parameters for the HMM, HaMMLET performs a number of Gibbs sampling steps. Starting from a prior sampling of parameters, which it can determine automatically, it iterates between sampling of possible segmentations (latent state sequences) conditioned on the previous parameters, and sampling of parameters conditioned on the previously sampled state sequence. After a certain number of ***burn-in*** iterations, the sampled state sequences are used to approximate the marginal probabilities of each CNV state at each position. This is then used to derive a maximum posterior margin (MPM) segmentation of the data.

2 Materials

2.1 Building HaMMLET

HaMMLET is written in C++ and requires compilation using a C++11-compliant compiler. Beyond the Standard Template Library (STL), which is typically included with any modern compiler, HaMMLET does not depend on any external software packages, libraries or runtime environments. It has been tested only in a Linux environment using g++. To install HaMMLET in a Linux environment, use

```
git clone https://github.com/wiedenhoeft/HaMMLET.git
cd HaMMLET
make
```

This creates an executable called "hammlet" in the bin/ subdirectory; for plotting as well as transforming WGS data into a form suitable for HaMMLET, we also provide a number of scripts and executables. To make the executables available, set

```
export PATH=$PATH:/your/directory/path/to/HaMMLET/bin/
```

We assume a reasonably modern Linux environment, with bash, gzip, sed, and awk installed, as well as Python 2.7 with Matplotlib. For working with WGS data, we assume that SAMtools is installed; for details, see https://github.com/samtools/samtools.

The doc/ subdirectory contains extensive documentation of HaMMLET in various formats (PDF, HTML, man, ...). In this chapter, we discuss only those options relevant to WGS segmentation.

3 Methods

HaMMLET assumes that the input data was generated by an HMM with Gaussian emission distributions. Beyond that, it is oblivious to the underlying nature of the data, and simply processes an array of numeric values, similar to software like CBS [10], and can therefore be used as a generic segmentation method for different experimental platforms. Using the appropriate best-practice preprocessing methods such as CG bias removal is advised if the number of CNV states provided to HaMMLET by the user is to represent that of the assumed number of actual CNV states in the data. Alternatively, HaMMLET itself can be run with a larger number of states to provide a segmentation which includes such bias, and then use downstream methods like clustering by means to obtain CNV calls.

In what follows, we provide an example of how to use read-depth information from multiplexed high-throughput sequencing as an input for HaMMLET. A simple pipeline for CNV calls in differential WGS data would be:

```
# Create count data for sample and control
  samToCounts sample.bam sample
  samToCounts control.bam control
# Combine into differential count data
  combineCounts -i + sample - control -o combined
# Average over 20 input positions
  gunzip -c combined-count.csv.gz | avg 20 | gzip -c
  > result-count.csv.gz
# Run HaMMLET
  gunzip -c result-count.csv.gz | hammlet -s 8 -a -o
  result- .csv -R 0 -O M P -v
# Sort states by absolute value of emission means
  sortStates result-parameters.csv > result-sorted_
  states.csv
# Map HaMMLET's segments back to the genome
  mapLinesToGenome -i result-marginals.csv -g com-
  bined -w 20 -r -b -c > result-regions.csv
# Create maximum posterior margins segmentation
  maxSegmentation -i result-marginals.csv > result-
  segmentation.csv
# Map MPM segmentation to genome and split files by state
  mapLinesToGenome -i result-segmentation.csv -g com-
  bined -w 20 -r -b -c | awk '{print $1 >> "result-
  state"$2".csv"}'
# Plot results
  plotResults -f result-count.csv.gz -i result- .csv -S 10000
```

Depending on the size of the SAM or BAM files and the disk speed, extracting the read counts with SAMtools can take several hours. HaMMLET itself usually only takes minutes even for WGS data. One should conservatively estimate 1.5 GB RAM for every 100 million values in result-count.csv.gz (*see* **Note 1**), and hence HaMMLET is unlikely to exceed the memory provided by modern consumer hardware (desktop or laptop). Plotting relies on Python's Matplotlib library and is very slow.

3.1 Preparing WGS Data

In multiplexed sequencing, individual barcodes are used to tag different samples, such as healthy vs. cancerous tissue, in the same sequencing run. Since both samples can be expected to share the same experimental bias, subtracting read-depth information of one sample to the other is expected to cancel out most of it. We assume that the data has been demultiplexed into two separate BAM files, e.g., sample.bam and control.bam. Being a generic method, HaMMLET operates on a simple array of input values, and is oblivious to genome coordinates. We provide some mechanisms to map the results back to genome coordinates. For a given file PREFIX, read-depth data is stored across three files:

- *PREFIX-size.csv* contains three tab-separated columns: the first column contains the names of reference sequences, the second column contains the number of genomic positions for each reference sequence for which counts have been observed, and the third column contains the cumulative sums of the second column.
- *PREFIX-pos.csv.gz* is a gzipped text file, containing the one-based genome coordinates for each reference sequence, in the same order as they appear in PREFIX-size.csv. The second column in PREFIX-size.csv denotes the number of lines in PREFIX-pos.csv.gz belonging to each refseq.
- *PREFIX-count.csv.gz* contains the observed counts or similar intensity data for each position, and the lines correspond to those in PREFIX-pos.csv.gz. For aCGH applications, this file would contain the log-intensity ratios. Corrections such as CG bias removal should be performed on this file.

Since HMM assume that the individual observed data points are conditionally independent given the true sequence of state labels, read-depth as obtained from pileup cannot be used directly, since it creates statistical dependencies between coverage of neighboring positions. Instead, the number of reads which have their left-most base mapped to any given position is recorded for both samples and then subtracted, creating a differential signal similar to that of aCGH. To create the individual counts from an input BAM or SAM file, use the shell script

```
samToCounts input.bam output
```

which creates output-size.csv, output-count.csv.gz and output-pos.csv.gz, as described above. A number representing SAM flags for filtering can be provided as a third argument, so that an alignment is ignored if any of its corresponding bits are set:

```
samToCounts input.bam output 3844
```

By default, 3844 is used, meaning that the following alignments are ignored: "read unmapped" (4), "not primary alignment" (256), "read fails platform/vendor quality checks" (512), "read is PCR or optical duplicate" (1024), or "supplementary alignment" (2048). Multiple alignments of the same read to the same position are ignored as alternative alignments.

In order to obtain differential line counts, we assume that the files sample.bam and control.bam have been processed using samToCounts. The counts are combined using the prefixes of those filenames (*see* **Note 2**):

```
combineCounts -i + sample - control -o combined
```

(notice the + and - passed as argument to -i). This subtracts control counts from the corresponding positions in the sample if such counts exist. If a position does not exist in the sample, it is implicitly assumed to be zero. The output consists of the same three file formats as before, using -o as a filename prefix. Multiple input files can be combined using

```
combineCounts -i + sample1 sample2 sample3 - control1
control2 control3 -o combined
```

which creates combined-count.csv.gz, combined-pos.csv.gz and combined-size.csv.

While HaMMLET will yield some segmentation of the combined count data, the noise is not Gaussian. To make the noise more normal, the data is binned and averaged. Using bins of size 20 is usually sufficient:

```
gunzip -c combined-count.csv.gz | avg 20 | gzip -c >
result-count.csv.gz
```

3.2 Running HaMMLET

The simplest way to run HaMMLET for a model with for example eight states is:

```
gunzip -c result-count.csv.gz | hammlet -s 8 -a -o result-
.csv -R 0 -O M P -v
```

Here, -a denotes that automatic priors should be used, -s specifies the number of states, and -o specifies the prefix and suffix for the output files (notice the space!). HaMMLET does not support reading gzipped input directly. For uncompressed input files, -f can be used instead of the pipe:

```
hammlet -f result-count.csv -s 8 -a -o result- .csv -R
0 -O M P -v -w
```

The -O option determines which output files are created by HaMMLET (*see* Table 1); the option is described in Section 3.4.

3.3 Sampling Schemes

Bayesian approaches require a repeated resampling of the latent state sequence, that is, many samples of possible segmentations. HaMMLET's provides a versatile mechanism, defined by a number of tokens to its -i option(See Table 2). It supports two basic sampling modes: *Forward-backward sampling (FBG)* is the standard approach to sampling state sequences (segmentations) in a hidden Markov model, leading to high-quality segmentation. It is a very memory-intensive algorithm in that it creates a large matrix called a *trellis*, whose size depends on the number of data points and the

Table 1
HaMMLET command line arguments for output

Argument	Value	Description	Default
-f -file	String	Path to input file	By default, the data is read from the console through a pipe (STDIN)
-o -output-pattern	Two strings	The prefix and suffix for the output file path	If *-f filename.ext* is provided, the default is *filename- .ext*, otherwise it is *hammlet- .csv*
-O -output-data	List of strings.	A list containing output formats (see section 3.4 below)	*marginals*
-w -overwrite		Overwrite existing files	By default, existing files will not be overwritten

number of HMM states. Compression reduces its size by the compression ratio, which can still leave a considerable number of values. FBG is requested using

```
-i F 100 3
```

where the first integer (100) stands for the number of sampling iterations, and the second number (3) stands for the thinning, meaning that, in this case, every third sequence is recorded (0 means that none of the sequences are recorded, which is useful for burn-in steps).

Mixture sampling on the other hand ignores all state transition probabilities and hence does not require a trellis. It is therefore very fast and uses almost no additional memory, but is unreliable for segmentation, as it tends to oversegment high-variance regions and switches state labels excessively. It is, however, useful during the initial stages of sampling, to arrive at a sensible compression ratio. Mixture sampling is requested in an analogous way, using

```
-i M 100 3
Aside from these triples of tokens, –i also accepts
the following single-letter arguments. By default, the
compression is dynamic and depends on previously sam-
pled parameters. By using
```

```
-i S
```

the compression can be fixed to the last sampled value. Likewise,

```
-i D
```

sets the compression to dynamic again. Lastly,

```
-i P
```

triggers sampling of HMM parameters from a prior. This step is also implicitly performed before the very first sampling iteration. HaMMLET's standard argument for the sampling scheme is

Table 2
Important HaMMLET command line arguments for the sampling model

Argument	Value	Description	Default
-s -states	Integer	The number of HMM states (more complicated state definitions exist for multivariate data, see manual)	3
-a -auto-priors	Pair of floats	Use automatic priors. The first argument defines a noise variance, the second a probability to sample emission parameters of at most that variance.	0.2 0.9
-t -transitions	Pair of floats	Parameters to the Dirichlet prior used to sample transitions. The first one is for transitions between states, the second for self-transitions.	0.5 0.5 (corresponds to Jeffreys prior)
-I -initial-dist	Float	Parameter to the Dirichlet prior used to sample the initial state distribution.	0.5 (corresponds to Jeffreys prior)
-R -random-seed	Integer	The seed for the random number generator. This value should be set by the user for reproducibility.	The current UNIX timestamp

```
-i M 500 0 S P F 200 0 F 300 3
```

This means that, after an implicit sampling of the model parameters from their prior, 500 unrecorded mixture iterations are performed. This typically leads to a good block structure, which is then fixed using S. Then, the parameters are sampled from the priors using P, in order to forget the unreliable emission parameters for the mixture model. Then, 200 unrecorded FBG steps are performed as a burn-in, followed by 300 FBG steps, every third of which is recorded, leading to a total of 100 recorded segmentations.

3.4 HaMMLET's Output Formats

The -O option is used to specify the output HaMMLET produces using single capital letters or individual words. The latter is used to denote the output file name, for instance, using

```
-o wgsdata_ .txt -O M
```

produces a file called wgsdata_marginals.txt. The following options are relevant for calling CNV from univariate data:

- *M/marginals*: State marginals are output as a tab-separated file in which the first column contains the segment size, and the following columns contain the counts for individual states. In each line, the state with the highest count is the state assigned to this segment by MPM segmentation.
- *S/sequences*: State sequences are written in run-length encoded format, segmentsize1:state1 segmentsize2:state2 ... with one state sequence per sampling iteration.
- *B/blocks*: The block structure is output as space-separated integers describing the number of data positions per block, with

one line for each sampling iteration. Notice that these are the block sizes used for compression before the sampling of a state sequence, whereas the segment sizes above are output after the sampling, so each segment can consist of several blocks.

- *C/compression*: Output the compression ratio, i.e., the number of data positions divided by the number of blocks, for each iteration.
- *P/parameters*: Output the emission parameters of the individual states at each iteration, in the form mean1 variance1 mean2 variance2 ...

3.5 Sorting States to Rank CNV Candidates

In the pipeline above, HaMMLET outputs the marginal state counts to result-marginals.csv, as well as the sampled emission parameters to result-parameters.csv. Since we use differential data, we assume that the segments which have the same copy number in sample and control will have a mean close to zero. Using

```
sortStates result-parameters.csv > result-sorted_states.
csv
```

takes the last line in result-parameters.csv, extracts the means and creates an output file in which the state labels (first column) are sorted in decreasing order by their absolute deviation from zero. States higher on the list are therefore better candidates for CNV than those at the bottom.

3.6 Mapping Segmentation Results to Genome Coordinates

The mapLinesToGenome tool was designed to map lines in a text file to genomic coordinates. To map the segmentation results obtained by HaMMLET, use.

```
mapLinesToGenome -i result-marginals.csv -g combined
-w 20 -r -b -c > result-regions.csv
```

This takes result-marginals.csv as input, and maps it against the reference files defined by the file prefix after the -g flag, i.e., combined-size.csv and combined-pos.csv.gz. The -w option specifies the number of genome positions to be associated with one line in the input file; in this case, this is the size of the window used for averaging. -r combines subsequent genome coordinates into range:

```
chr2   5577   6013   0   29   14   12   19   13   0   13
```

where chr2 is the refseq, 5577 is the 1-based start and 6013 is the inclusive end position. Using -c separates these ranges by colon rather than tabs:

```
chr2:   5577:   6013   0   29   14   12   19   13   0   13
```

If -b is provided, the first tab-separated column must be an integer and is treated as run-length encoding. This fits the output format of state marginals, where the first column describes the size of the segment in terms of the number of input positions. For instance,

```
256     0     29     14     12     19     13     0   13
```

means that 256 consecutive data points in the input file have the same marginal counts as specified by the remaining columns. Notice that this column is removed by mapLinesToGenome -b, and the results are in the same format as above.

While the individual state counts in the marginals are interesting in their own right, in many cases one is particularly interested in the maximum posterior margins segmentation. This can be achieved using

```
maxSegmentation -i result-marginals.csv > result-
segmentation.csv
```

which transforms

```
256     0     29     14     12     19     13     0   13
```

into

```
256 1
```

since state 1 has the highest number of observations (29). Then,

```
mapLinesToGenome -i result-segmentation.csv -g com-
bined -w 20 -r -b -c | awk '{print $1 >> "result-
state"$2".csv"}'
```

maps those lines to genome coordinates, thereby associating each input position with is maximum marginal state. This information is then used to split the result into files named result-state#.csv, where # is the state label. Those files contain colon-separated genome coordinates. Then, result-sorted_states.csv can be used to choose a subset of these files as putative CNV for downstream analysis such as GO term enrichment etc.

3.7 Plotting

HaMMLET comes with a Python script called plotResults (*see* **Note 3**). Using

```
plotResults -f result-count.csv.gz -i result- .csv -S 10000
```

produces figures displaying the data, MPM segmentation as well as the state marginals, see Fig. 1 for an example. The -f option describes the input data file, the -i option describes the prefix and suffix for supporting data such as marginal counts (here: result-marginals.csv); other data such as block sizes can be plotted as well, see

```
plotResults -h
```

for more information. The -S option limits the number of data points for each figure to 10,000. Using too high a number here

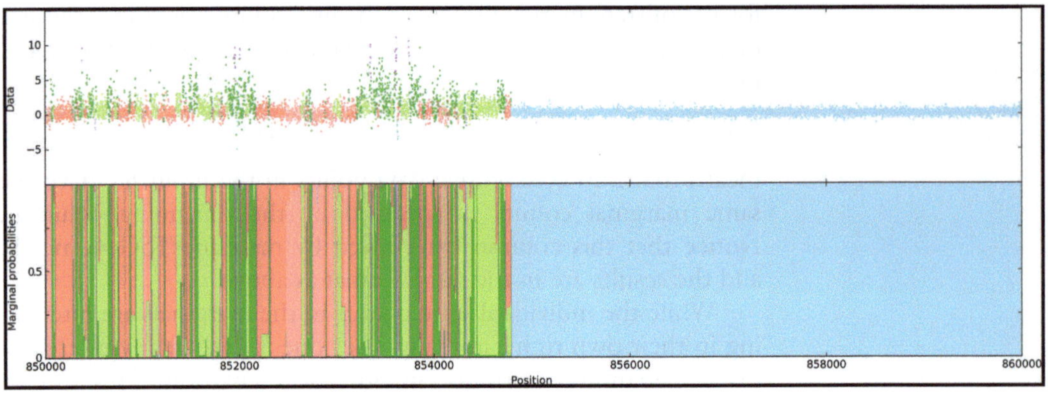

Fig. 1 Example output from the auxillary script "plotResults" included in the HaMMLET package. The plot displays the data range (top) and color-coded, state, marginal probabilities (bottom) for each CNV region

can lead to excessive memory loss and program crashes, since the Matplotlib library does not appear to be optimized for huge data (*see* **Note 4**).

4 Notes

1. While HaMMLET has been carefully designed to minimize memory consumption, one should still be aware that the memory requirements cannot reliably be determined beforehand. Data may contain vast numbers of complex CNV, leading to low compression ratios which can incur huge memory overhead, as such data negates HaMMLET's central approach of using Haar wavelets that makes FBG feasible on such scales. This is particularly true for the forward-backward sampling steps, since they require an additional trellis data structure whose size depends on the number of states as well as the number of compression blocks. However, such data is likely to be an exceptional case. If memory consumption does get out of hand, one can try increasing the number of burn-in steps; if the sampler has not fully converged, individual iterations might have very low compression, even though the data itself would allow for better ratios. Likewise, decreasing the number of states might be an option, since superfluous state parameters will be sampled solely from the prior and yield arbitrarily low noise variances. Finally, using unrecorded mixture sampling with dynamic compression, then switching to static compression for FBG is a strategy which often works. It should be emphasized that the last sampling steps should always be FBG; while mixture sampling helps in increasing compression during burn-in, the segmentation it yields is unreliable!

2. If the default arguments to -i are used, FBG is sampled only on a static compression of the data. If the resulting segmentation seems wrong, the sampling scheme should be changed to include dynamic compression for FBG, at the risk of incurring higher memory consumption.
3. Plotting the results is done using external Python libraries (NumPy, Matplotlib). As these are not optimized for large-scale applications, this can take a long time, often much longer than the inference itself.
4. HaMMLET does not support the convention of stacking single-letter options, such as replacing -x -y -z by -xyz.

References

1. Wiedenhoeft J, Brugel E, Schliep A (2016) Fast Bayesian inference of copy number variants using hidden Markov models with wavelet compression. PLoS Comput Biol 12(5):e1004871. https://doi.org/10.1371/journal.pcbi.1004871
2. Bilmes JA (1998) A gentle tutorial of the EM algorithm and its application to parameter estimation for Gaussian mixture and hidden Markov models. Int Comput Sci Inst 4(510):126 Retrieved from http://citeseerx.ist.psu.edu/viewdoc/summary?doi=10.1.1.28.613
3. Rabiner L, Juang B (1986) An introduction to hidden Markov models. IEEE ASSP Mag 3(1):4–16. https://doi.org/10.1109/MASSP.1986.1165342
4. Rabiner LRLR (1989) A tutorial on hidden Markov models and selected applications in speech recognition. Proc IEEE 77(2):257–286. https://doi.org/10.1109/5.18626
5. Chib S (1996) Calculating posterior distributions and modal estimates in Markov mixture models. J Econ 75(1):79–97 Retrieved from http://www.sciencedirect.com/science/article/pii/0304407695017704
6. Scott SL (2002) Bayesian methods for hidden Markov models: recursive computing in the 21st century. J Am Stat Assoc 97(457):337–351. https://doi.org/10.1198/016214502753479464
7. Shah SP, Xuan X, DeLeeuw RJ et al (2006) Integrating copy number polymorphisms into array CGH analysis using a robust HMM. Bioinformatics 22(14):e431–e439. https://doi.org/10.1093/bioinformatics/btl238
8. Shah SP, Lam WL, Ng RT et al (2007) Modeling recurrent DNA copy number alterations in array CGH data. Bioinformatics 23(13):i450–i458. https://doi.org/10.1093/bioinformatics/btm221
9. Guha S, Li Y, Neuberg D (2006) Bayesian hidden Markov modeling of Array CGH data, Harvard University Biostatistics Working Paper Series. Retrieved from http://biostats.bepress.com/harvardbiostat/paper24
10. Olshen AB, Venkatraman ES, Lucito R et al (2004) Circular binary segmentation for the analysis of array-based DNA copy number data. Biostatistics (Oxford England) 5(4):557–572. https://doi.org/10.1093/biostatistics/kxh008

Chapter 7

Split-Read Indel and Structural Variant Calling Using *PINDEL*

Kai Ye, Li Guo, Xiaofei Yang, Eric-Wubbo Lamijer, Keiran Raine, and Zemin Ning

Abstract

Genetic variations are important evolutionary forces in all forms of life in nature. Accurate and efficient detection of various forms of genetic variants is crucial for understanding cell function, evolution and diseases in living organisms. In this chapter, we describe a detailed protocol that uses *Pindel*, a split-read algorithm, to discover indels and structural variants in a given genome, from Illumina short-read sequencing data produced from biological samples.

Key words PINDEL, Split-read, Structural variants

1 Introduction

Genetic variants are nucleotide changes occurred in genomes in the form of substitutions, small insertions and deletions (indels) and large complex structural variants. The most abundant genomic variations are single or multiple consecutive nucleotide substitutions. With a higher density than other types of variants, single nucleotide polymorphisms (SNPs) are frequently utilized in genome-wide association studies to pinpoint genome fragments linked to diseases or other traits. Besides substitutions, insertions or deletions of DNA bases are another important source of genomic variations between individuals. Structural variation (SV), originally defined as variation other than substitutions and small indels, ranges from large insertions, duplications, deletions to inversions and translocations of DNA fragment of 1 kb length or larger size [1]. As sequencing technologies and analytic algorithms advance quickly in recent years, the definition of SV has been expanded to include variants as small as 50-bp long [2–4]. In this protocol, indels are defined as insertions and deletions smaller or equal to 50 bp and SV as structural variants longer than 50 bp.

Derek M. Bickhart (ed.), *Copy Number Variants: Methods and Protocols*, Methods in Molecular Biology, vol. 1833, https://doi.org/10.1007/978-1-4939-8666-8_7,

Over the recent years, sequencing technology has propelled genomic studies and numerous computational software have been developed to deepen our understanding of structural variation and its role in genome functions [5–7]. Rapid depluming of NGS costs also makes genome-wide SV detection feasible, even at population scale [2–4]. Detection of structural variants is an active research area and different computational methods are being explored with various applications [8]. Generally, the algorithms used by these methods can be divided into four different categories: read-pair (RP), split-read (SR), read-depth (RD), and assembly (AS). RP detects SV based on the fact that mapping distance between two reads in a pair or orientation of either read is abnormal. As name suggests, RD algorithm detects SV based on the logic that duplicated or deleted regions tend to have significantly higher or lower coverage (read-mapping density) in sample compared to reference genome. AS is an approach that performs local assembly on read sets and call variants from the assembled contigs, under assistance of a reference genome. While a number of methods utilize only one of the above algorithms for SV detection, several recent published tools integrate two or more of these algorithms in order to elevate specificity and sensitivity [9, 10].

The focus of this protocol, *Pindel*, is a powerful indel and SV detection tool mainly deploying the SR (split-read) algorithm [11], the principle of using SR to detect variants is show in Fig. 1. Similar to read-pair signals, a structural variant often leaves a split-read signal in short read data, located at the breakpoints of the variant. SR algorithm identifies SV by splitting reads at the SV breakpoints during mapping processes, because the mapping location and orientation of the split-mapped reads will pinpoint the precise location, size and types of variants. *Pindel* is often referred to as anchored split-read mapping. When reads were short, *Pindel* uses one of the reads as an anchor in order to reduce the search space of the split-mapping of its mate. Pattern growth, originally a sequential pattern mining approach, was used in the package for efficient string matching. Compared to other approaches, SR approach, such as *Pindel*, is powerful for detecting small and medium-size variants such as insertions, deletions, tandem duplications, and inversions, as well as translocations, providing precise breakpoints with base accuracy down to single base levels. SR has limitations for detecting large variants or those in repetitive regions due to its local mapping procedure.

2 Materials

2.1 Hardware

The computational requirement for *Pindel* is mainly the memory storage, which depends on the size of the genome, the depth of sequencing, the length of reads and the size of window

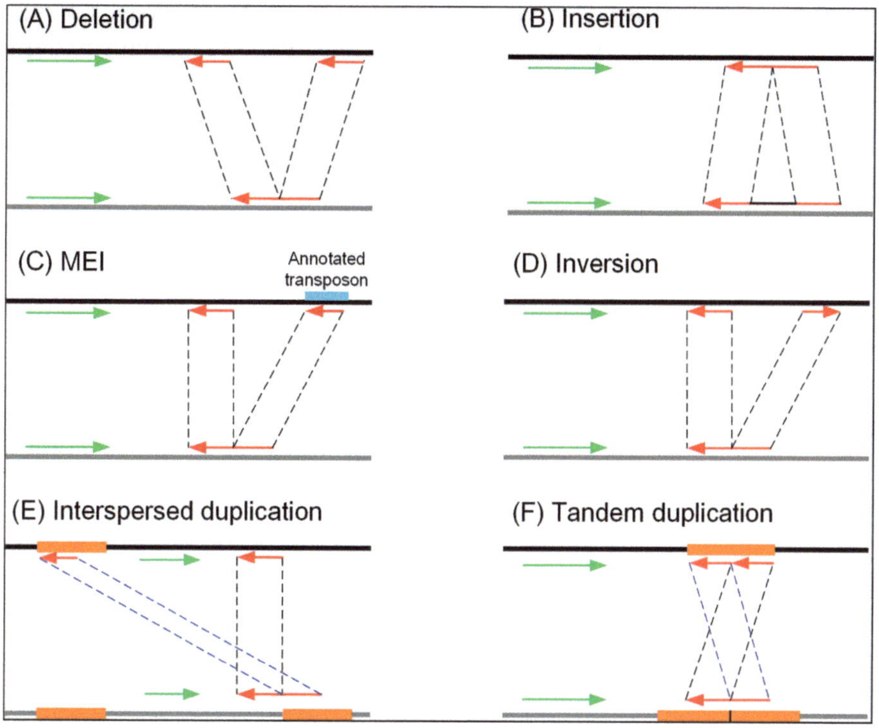

Fig. 1 Principle of using split-read to detect variants. The black lines represent reference genome and the gray lines represent sample genome; the green arrows represent the anchors; the red arrows represent the split reads (SR); the orange blocks represent the duplicated DNA sequences; and the dash lines represent the SR mappings. *MEI*: mobile element insertion

(detail describe below). For instance, we tested the *Pindel* on chromosome 20 of NA12878 sample (data was downloaded from http://smash.cs.berkeley.edu/datasets.html) with the sequence depth 53×. We found that it required ~583 MB memory and 13 min to run *Pindel* with the default parameters (*see* **Note 1**) using *Pindel* version 0.2.5b9 on a computer with Intel(R) Core(TM) i5-4460 CPU @3.20 GHz) .

In addition, to suit different types of hardware, there are two parameters that users can set to affect the runtime and memory usage (*see* **Note 2**):

-T/--number_of_threads	The number of threads *Pindel* will use (default 1). More threads assure lower runtime, but requires multiple processors (*see* **Note 3**)
-w/--window_size	For saving RAM, it divides the reference in bins of X million bases and only analyzes the reads per bin (default 10 (=10 million)). A smaller bin size will reduce memory but will increase runtime slightly

2.2 Software

Pindel is developed for Linux or Max OS in C++ language. Additionally, there are some functions are implemented using Perl, such as "bam2*Pindel*" function, which extracts *Pindel* format one end mapped reads from BAM files produced by BWA on Illumina reads. In this case, both GCC compiler and Perl environment should be also installed in your computer. There are two software packages on which *Pindel* depends:

Samtools [12]: a software that provides various utilities for manipulating alignments in the SAM/BAM format; the home page is at http://samtools.sourceforge.net/.

htslib: an implementation of a unified C library for accessing common file formats, such as SAM, CRAM, and VCF, used for high-throughput sequencing data, and is the core library used by samtools, the project page is at https://github.com/samtools/htslib.

3 Method

3.1 Download and Compile Pindel

Pindel source code is available on Github and can be downloaded easily using git clone command:

git clone https://github.com/genome/Pindel

To compile *Pindel* you need three things: GNU Make and GCC (which usually are already installed on Linux) and *htslib*. The last is not installed on Linux by default, but it can be retrieved with:

git clone https://github.com/samtools/htslib

htslib needs to be built before you can start installing *Pindel*. It simply works if you give the commands "make" and "sudo make install". If *htslib* has been cloned and installed, go to the *Pindel* directory ([my-path]/*Pindel*) and use the INSTALL script there in the following way:

./INSTALL *[path-to-htslib]*

To test whether *Pindel* has been properly installed, simply type "./*Pindel*" after prompt in terminal without giving any arguments or options. A help page printed out on the screen means a successful installation of *Pindel* and it's ready to use.

3.2 Running Pindel for Variant Discovery

The workflow for running Pindel to discovery different types of variants is show in Fig. 2. The details is described in the following section.

3.2.1 Preparing Input Data for Pindel from Read Alignment Data

Hereafter, we will describe in details how you can use *Pindel* to discover indels and structural variants from NGS data. A very simple example with four reads that support a deletion event is used to go into more detail. For genetic variant discovery in resequencing

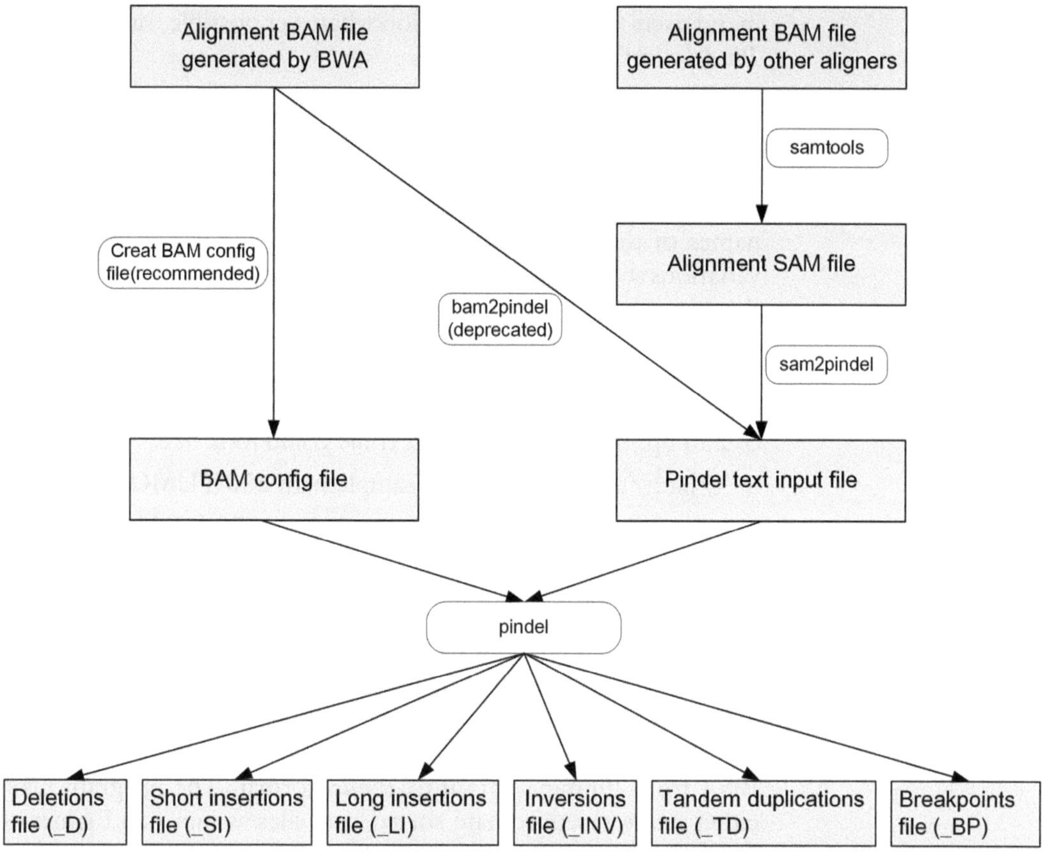

Fig. 2 The workflow for running *Pindel* to discover various variants. The rectangles represent the input or output files of *Pindel*, and the rounded rectangles represent the software packages or options users need to run

projects, typically DNA sequencing reads, if passing given quality thresholds, are aligned to a reference genome sequence at hands using tools such as *BWA*, *Bowtie*, or *MOSAIK*, to produce a read alignment file in SAM/BAM format. For this protocol, we assume such alignment step has been completed by users already so that the alignment data are ready to run variant discovery using *Pindel*. To produce such read alignment files, raw sequence reads data can be processed using the following recommended protocol described elsewhere (https://github.com/lh3/bwa).

Before running *Pindel* on alignment data, useful reads need to be extracted from alignment files in three ways (Fig. 2): (1) indexed BAM files produced by BWA or MOSAIK on illumine reads, via a configuration file; (2) *Pindel* format reads extracted using *bam2Pindel.pl* from BAM files produced by BWA on Illumina reads; (3) reads prepared from SAM/BAM files not generated by BWA. Here we provide detailed instructions on how to use each of the three listed options to extract reads for *Pindel* run. However, we recommend users to use the first option whenever possible, since it makes *Pindel* run faster.

Frist option (recommended): Indexed BAM files produced by BWA or MOSAIK on Illumina reads, via a configuration file. *Pindel* version 0.2.0 and above can directly read in BWA/MOSAIK-generated BAM-files. To use this option, a small text-file containing the names of the BAM-files used to search for indels and structural variations should be created. Within the text-file, each line contains the name of the bam file, the insert size, and the label for the sample. For example, given two bam files named "tumor_sample.bam" and "somatic_sample.bam" with insert sizes (the length of sequence between the paired-end adapters in paired-end sequence) of 250 bp, the configuration text file could look like:

/path/to/folder/tumor_sample.bam 250 TUMOR

/path/to/folder/normal_sample.bam 250 HEALTHY

When running *Pindel*, use the -i option followed by the name of the config file.

Second option: *Pindel* format reads extracted using bam2*Pindel*.pl from BAM files produced by BWA on Illumina read. The program bam2*Pindel* was developed by Keiran Raine at the Wellcome Trust Sanger Institute (UK), designed to convert BAM files processed by BWA from Illumina data into *Pindel* format. The program takes either name or coordinate sorted bam files as input and generate *Pindel* format files as output. As of *Pindel* version 0.2.0 and above, this option of using bam2*Pindel*.pl is no longer the default running mode, as the configuration file option mentioned above saves disk space. However, it is still possible to use bam2*Pindel*. If this option is chosen, these steps should be followed:

Step 1. Download the BWA_BAM_2_*PINDEL*.tar.gz package from one of the following sources, and save it in a directory of your choice (say:'/user/*Pindel*-0.2.0/'):

https://trac.nbic.nl/*Pindel*/downloads

http://www.ebi.ac.uk/~kye/*Pindel*/v_0.2.0/BWA_BAM_2_*PINDEL*.tar.gz

ftp://ftp.sanger.ac.uk/pub/zn1/*Pindel*/*Pindel*_0.2.0.tar.gz

Unzipping the BWA_BAM_2_*PINDEL*.tar.gz will produce four files: the main perl script *bam2Pindel.pl*, a perl module to configure the tool called *Adaptor.pm*, a binary program to locate reads in a pair called *samgroupbyname*, and lastly a README file containing instructions how to use the tool.

Step 2: Set BAM_2_*PINDEL*_ADAPT in the way that your shell prescribes (you can find your shell type with "echo $SHELL"):
in csh:

setenv BAM_2_PINDEL_ADAPT /user/Pindel-0.2.0/Adaptor.pm

in bash, ksh, and sh:

export BAM_2_PINDEL_ADAPT=/user/Pindel-0.2.0/Adaptor.pm

Step 3: Run *bam2Pindel* on the input file, for example:

./bam2Pindel.pl –i tumor_sample.bam –o tumor_Pindelformat -s tumor –nm –pi 250

Third option: Reads prepared from SAM/BAM-file not produced by BWA. If another aligner than BWA was used, the best way to convert the reads to *Pindel* input format is to use sam2*Pindel*.cpp. Sam2*Pindel* expects a mapping location of mate read if the query is unmapped while the mate has been mapped. Users may need to fix the information about mate reads for some mappers such as bowtie. Compiling and running sam2*Pindel*.cpp works as follows:

Step 1: Download the cpp code (SAM_2_*PINDEL*_cin_2010Dec2.cpp) from either of the following sources:

https://trac.nbic.nl/*Pindel*/downloads

http://www.ebi.ac.uk/~kye/*Pindel*/v_0.2.0/SAM_2_*PINDEL*_cin_2010Dec2.cpp

ftp://ftp.sanger.ac.uk/pub/zn1/*Pindel*/*Pindel*_0.2.0.tar.gz

Step 2: Compile the cpp source code:

g++ SAM_2_*PINDEL*_cin_2010Dec2.cpp –o sam2*Pindel* –O3

Step 3: Run *sam2Pindel*. If the input is a SAM file, the command should look similar to this:

./sam2*Pindel* input.sam Output4*Pindel*.txt 300 tumour 0

If the input is a BAM file, *sam2Pindel* should be run like this:

./samtools view input.bam | ./sam2*Pindel* - Output4*Pindel*.txt 300 tumour 0

sam2Pindel requires five arguments: Input sam file (here “input.sam”), Output for *Pindel* (here “Output4*Pindel*”), insert size (here 300): the length of sequence between the paired-end adapters in paired-end sequence sample tag: (here “tumor”), number of extra lines (nonrecord and not starting with @) in the beginning of the file to skip (here 0).

3.3 Calling Indels and Structure Variants

If a BWA-BAM file and a BAM configuration file are available using the first option, to discover indels and structural variants from the data, one simply runs *Pindel* with following basic parameters: -f to specify the reference fasta file, -i to specify the input configuration file created, -c to specify the chromosome or fragments, and –o to specify the output filename. For example:

./Pindel -f hs_ref_GRCh37.fa -i config.txt -c ALL -o sample

This command line will allow *Pindel* to find indels and SV in the alignment data files specified in config.txt in all chromosomes. The output files should then contain all detected indels and SVs relative to the reference, and the labels will indicate in which samples each indel and SV occurred. For example, sample_D, sample_SI, sample_LI, sample_INV, sample_TD contains the deletions, short insertions, long insertions, inversions, and tandem duplications found in the sample against the reference genome.

Typically genetic variants are reported in a specific file format called VCF (variant call format) [13]. *Pindel* itself does not output VCF files directly. Instead, it outputs a file format that gives far more details on the alignments of the reads supporting the variants. However, *Pindel* package does provide a conversion utility called *Pindel2vcf* that converts standard *Pindel* output files to VCF format. This utility can be found as an executable file in the same directory as *Pindel* itself. It can be used as following (*see* **Note 4**):

./Pindel2vcf –p sample_D –r hs_ref_GRCh37.fa –R GRCh37 –d 20121106 –v sample_D.vcf

Where the *–p* option is the name of the *Pindel* output file, *-r* the name of the reference fasta file, *-R* the official name of the reference, *-d* the date at which the reference was originally created.

If option 2 or 3 is used, one should first decide whether you want to create an output file that directly compares the indels and structural variations in two samples (for example, tumor versus normal), in that case, concatenate the files first (for example: "cat tumor.txt somatic.txt > total_sample.txt"). Run *Pindel* with the -p option followed by the name of the *Pindel* input file, so

./Pindel -f hs_ref_GRCh37.fa -p tumor_chr1.txt -c ALL -o tumor

4 Notes

1. *Pindel* has many parameters that you can set to increase speed (multithreading) or change the balance between sensitivity and specificity.

The basic parameters:

-f/--fasta	The reference genome sequences in fasta format
-p/--*Pindel*-file	The *Pindel* input file (either this or a bam configuration file is required)
-i/--config-file	The bam config file; either this or a *Pindel* input file is required. Per line: path and file name of bam, insert size, and sample tag. For example: /data/tumour.bam 400 tumour
-o/--output-prefix	Output prefix

-c/--chromosome	Which chr/fragment. *Pindel* will process reads for one chromosome each time. ChrName must be the same as in reference sequence and in read file. '-c ALL' will make *Pindel* loop over all chromosomes. The search for indels and SVs can also be limited to a specific region; -c 20:10,000,000 will only look for indels and SVs after position 10,000,000 == [10M, end], -c 20:5,000,000-15,000,000 will report indels in the range between and including the bases at position 5,000,000 and 15,000,000 = [5M, 15M]

Runtime and memory usage

-T/--number_of_threads	The number of threads *Pindel* will use (default 1). More threads assures lower runtime, but requires multiple processors
-w/--window_size	For saving RAM, divides the reference in bins of X million bases and only analyzes the reads per bin (default 10 (=10 million)). A smaller bin size will reduce memory but will increase runtime slightly

Structural variant types

-x/--max_range_index	The maximum size of structural variations to be detected; the higher this number, the greater the number of SVs reported, but the computational cost and memory requirements increase, as does the rate of false positives. 1=128, 2=512, 3=2,048, 4=8,092, 5=32,368, 6=129,472, 7=517,888, 8=2,071,552, 9=8,286,208 (maximum 9, default 5)
-r/--report_inversions	Report inversions (default true)
-t/--report_duplications	Report tandem duplications (default true)
-l/--report_long_insertions	Report insertions of which the full sequence cannot be deduced because of their length (default true)
-k/--report_breakpoints	Report breakpoints (default true)
-s/--report_close_mapped_reads	Report reads of which only one end (the one closest to the mapped read of the paired-end read) could be mapped (default false)
-n/--min_NT_size	Only report inserted (NT) sequences in deletions greater than this size (default 50)
-v/--min_inversion_size	Only report inversions greater than this number of bases (default 50)

Sensitivity and selectivity:

-d/--min_num_matched_bases	Only consider reads as evidence if they map with more than this number of bases to the reference (default 30)
-a/--additional_mismatch	*Pindel* will only map part of a read to the reference genome if there are no other candidate positions with no more than the specified number of mismatches position. The bigger this value, the more accurate but less sensitive (default value 1)

-m/--min_perfect_match_around_BP	At the point where the read is split into two, there should at least be this number of perfectly matching bases between read and reference (default value 3)
-e/--sequencing_error_rate:	The expected fraction of sequencing errors (default 0.05)
-u/--maximum_allowed_mismatch_rate:	Only reads with fewer mismatches with the reference genome than this fraction will be considered (default 0.1)

Miscellaneous parameters

-b/--breakdancer [file name]	*Pindel* is able to use calls from other SV methods such as BreakDancer to further increase sensitivity and specificity. BreakDancer result or calls from any methods must be in the format: ChrA LocA stringA ChrB LocB stringB other
-Q [file name]	The list of BreakDancer calls with *Pindel* support information. Format: chr Loc_left Loc_right sizetype index. For example, "172766323 72811840 45516 D 11970" means the deletion event chr1:72766323-72811840 of size 45516 is reported as an event with index 11970 in *Pindel* report of deletion
-h/--help	Show the command line options of *Pindel*

2. Running *Pindel* is not memory-intensive. However, it may have a high memory usage when trying to process the centromere regions, which contain lots of 'weird' reads. In such cases, it is best to use the -j or -J options to specify which chromosomal regions should be searched by *Pindel* (-j) or skipped (-J). Note that in some cases you may need to extend the excluded regions a bit (by 10 k or such), as *Pindel* by default also checks for reads just outside the edges of the officially specified regions, so you may want to slightly enlarge your centromeres, telomeres or other problematic regions. The window size option (-W) can also reduce memory usage for very large data sets.

3. *Pindel* can run quite slowly if there are many reads to process; for a part, long runtime can be caused by high coverage. However, there are some options to speed things up:

   (a) Eliminate 'unproductive' searching by excluding the centromeric and telomeric regions from the search process, this can be done with the -j and -J options.

   (b) Parallelize: there are two ways in which to speed up *Pindel* using parallelization:

   - Allow *Pindel* to use multithreading (multiple processors/processor cores)
   - by using the -T option (default is 1; setting it to 2, 4, or 6 may help, depending on your system)
   - Commit jobs per chromosome or half-chromosome in parallel to a cluster, this can be done by using the -c

option to specify a chromosome (like -c 20) or per chromosome region (-c 20:500000-2000000)

4. Despites its name, *Pindel2vcf4tcga* does not produce valid VCF files. This is because valid TCGA VCF files require a lot of input data that *Pindel2*vcf cannot know (like the version and settings of the alignment software). Even if all that data is available, the file produced would likely still be unacceptable to the TCGA database as the TCGA database only accepts files from "accepted" groups and projects, and regular *Pindel* users would therefore find *Pindel2vcf4tcga* of no use at all. For now, *Pindel2vcf4tcga* is basically a "half-converter" which converts *Pindel* files into a semi-TCGA-vcf format, which can be processed by certain scripts that TCGA partners possess into full-blown TCGA files. This does not mean that *Pindel2vcf4tcga* will never be working as a "stand-alone" script, but for now it is simply for use by certain TCGA-affiliated users, regular *Pindel* users should not try to apply it to their own *Pindel* output.

Acknowledgment

The authors gratefully acknowledge the financial support from General Program of National Natural Science Foundation of China (No. 31671372, 61702406).

References

1. Freeman JL, Perry GH, Feuk L et al (2006) Copy number variation: new insights in genome diversity. Genome Res 16:949–961
2. Mills RE, Walter K, Stewart C et al (2011) Mapping copy number variation by population-scale genome sequencing. Nature 470:59–65
3. Sudmant PH, Rausch T, Gardner EJ et al (2015) An integrated map of structural variation in 2,504 human genomes. Nature 526:75–81
4. Zarrei M, MacDonald JR, Merico D et al (2015) A copy number variation map of the human genome. Nat Rev Genet 16:172–183
5. Koboldt DC, Steinberg KM, Larson DE et al (2013) The next-generation sequencing revolution and its impact on genomics. Cell 155:27–38
6. Lin K, Smit S, Bonnema G et al (2015) Making the difference: integrating structural variation detection tools. Brief Bioinform 16:852–864
7. Tattini L, D'Aurizio R, Magi A (2015) Detection of genomic structural variants from next-generation sequencing data. Front Bioeng Biotechnol 3:92
8. Guan P, Sung WK (2016) Structural variation detection using next-generation sequencing data: a comparative technical review. Methods 102:36–49
9. Layer RM, Chiang C, Quinlan AR et al (2014) LUMPY: a probabilistic framework for structural variant discovery. Genome Biol 15:R84
10. Rausch T, Zichner T, Schlattl A et al (2012) DELLY: structural variant discovery by integrated paired-end and split-read analysis. Bioinformatics 28:i333–i339
11. Ye K, Schulz MH, Long Q, Apweiler R, Ning Z (2009) Pindel: a pattern growth approach to detect break points of large deletions and zshort reads. Bioinformatics 25:2865–2871
12. Li H, Handsaker B, Wysoker A et al (2009) The sequence alignment/map format and SAMtools. Bioinformatics 25:2078–2079
13. Danecek P, Auton A, Abecasis G et al (2011) The variant call format and VCFtools. Bioinformatics 27:2156–2158

Chapter 8

Detecting Small Inversions Using SRinversion

Ruoyan Chen, Yu Lung Lau, and Wanling Yang

Abstract

Rapid development of next generation sequencing (NGS) technology has substantially improved our ability to detect genomic variations. However, unlike other variations, such as point mutations, insertions, and deletions, which can be identified in high sensitivities and specificities based on NGS reads, most of inversions, especially those shorter than 1 kb, remain difficult to detect. Here we introduce a new framework, SRinversion, which was developed specifically for detection of inversions shorter than 1 kb by splitting and realigning poorly mapped or unmapped reads of the NGS data.

Key words Short inversion detection, NGS, Structural variations, Split reads method

1 Introduction

Inversions are a type of structural variations (SVs) where sequences are reversed compared with the reference genome. Previous studies indicate that inversions might play significant roles in population diversity and disease susceptibility [1–5]. In general, inversions play a functional role mainly in two ways: (1) one is directly associated with pathogenicity, such as the inversion that transforms structures of the factor VIII (F8) gene that is proved to be causal for hemophilia A [1], and the iduronate 2-sulfatase (IDS) gene that is associated with Hunter syndrome [2], (2) another is to have functional impact by interacting with other types of variants, such as the reported case where inherited inversions in patients activated one disease-associated copy number variation (CNV) [3–5]. However, although more and more studies have highlighted the important role of inversions in human diseases, efforts to detect this type of genetic variations has been lagging behind.

Among the large number of computational software for identification of SVs using NGS data, there are several of them that include functions for inversion detection [6–10]. However, only one [9] of these methods was developed specifically for identifying short inversions. Most of these tools detect inversions based on

Derek M. Bickhart (ed.), *Copy Number Variants: Methods and Protocols*, Methods in Molecular Biology, vol. 1833, https://doi.org/10.1007/978-1-4939-8666-8_8,

read pair information. In another words, they call inversions by detecting signals of abnormal directions of the reads in a read pair after aligning paired end NGS reads to a reference genome [6, 8]. And there are two prerequisites that must be fulfilled before identification of variations: (1) to be included for further analysis, read pairs must be able to be aligned to reference genome with quality score higher than cutoff set by the program, (2) at least one read of each read pair needs to cover the event to be identified, and the overlap between the read and the event should be long enough to be captured. As a result, inversions that are short enough to be covered by the entire read will be hardly identified. Since in these cases, most of the reads will be recorded as unmapped or low quality after alignment and will be discarded by these tools. Programs like Pindel [10] try to get rid of the limitation mentioned above by combining read pair information with split read methods, which makes it achieving higher sensitivity in inversion calling than most of the existing methods. However, prerequisites such as at least one of the "anchors" of a read pair need to be aligned to reference with high quality reducing its power in detecting one particular type of inversions: ones located near or within repetitive genomic regions. And as reported previously [11], inversions, especially smaller ones, tend to appear around complex regions resulted from their formation mechanism.

In this chapter we introduce SRinversion [12], a framework designed specific to detect inversions shorter than 1 kb based on NGS data. SRinversion applies an improved split read method. And instead of the traditional split read methods that relying on mapped reads as anchors, SRinversion makes use of information stored in poorly mapped or unmapped reads. By extracting, splitting or inverting, and remapping all of the low-quality or unmapped reads that are overlooked by most existing methods, the sensitivity of SRinversion is improved greatly.

In general, SRinversion takes four steps to call inversions: (1) extract poorly mapped reads, (2) split extracted reads from step 1, (3) realign split reads, and (4) figure out breakpoints of inversions.

For reads that have different physical distances with overlapped inversions, different split methods are required. So SRinversion applies two scenarios with respect to ways how candidate reads cover certain inversions (Fig. 1).

Fig. 1 (continued) part of each candidate read extracted in step 1 is inverted into its complementary sequence, and a sliding window with preset window size and step size is applied in order to screen all possible small inversions covered by candidate reads, then in step 3, all inverted reads are mapped to reference using single read model, finally in step 4, length and start position of inversions that are totally covered by candidate reads are figured out (reproduced from ref. 12 with permissions from Oxford University Press)

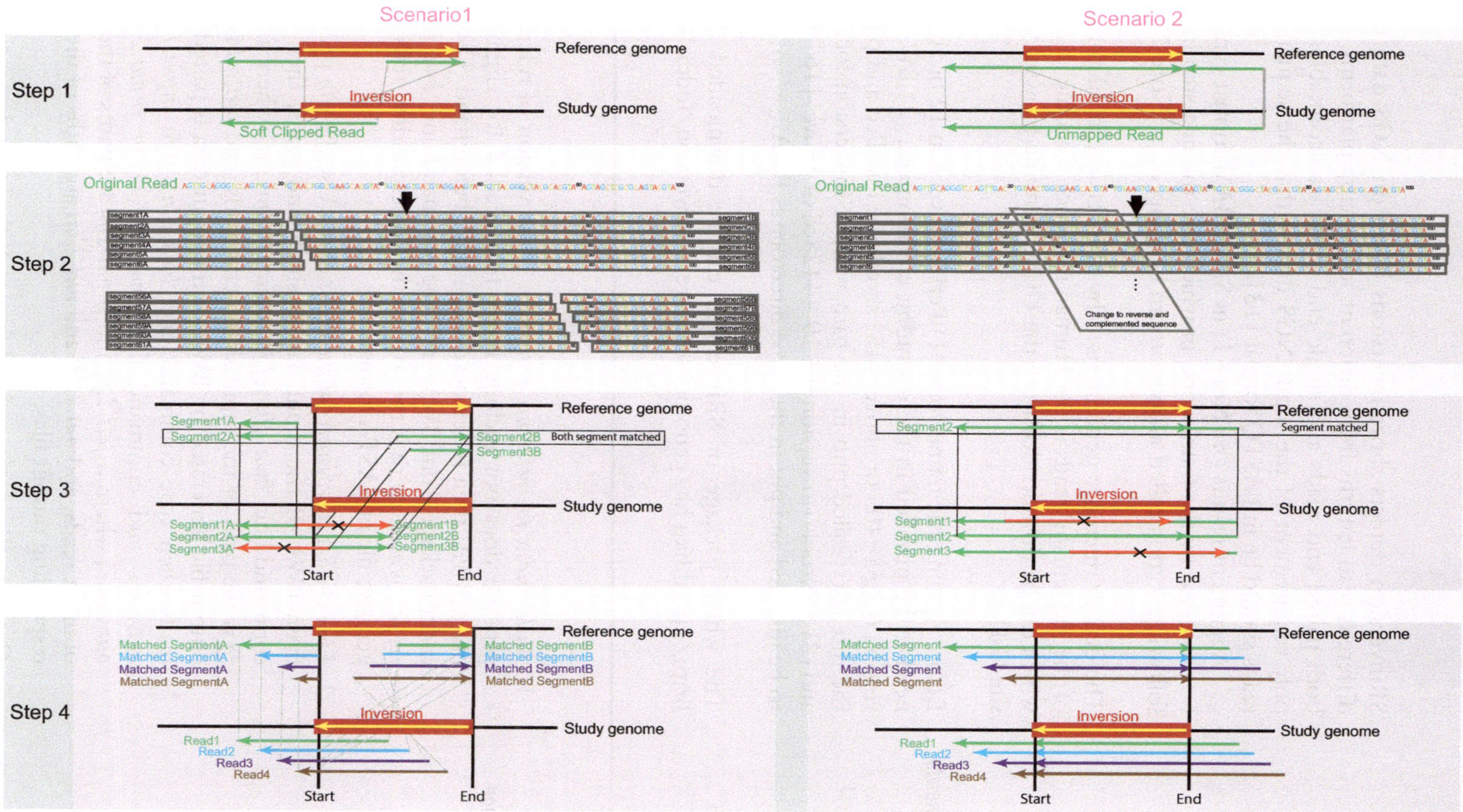

Fig. 1 Pipeline of SRinversion based on two scenarios. ***Scenario 1***: first in step 1, reads that are marked as soft clipped after first round of alignment were extracted from original bam file, secondly in step 2, each candidate read extracted from step 1 is split into several segment pairs at different position within the read, then in step 3, the split segment pairs were realigned to reference genome, finally in step 4, signals of all mapped segment pairs are integrated to figure out the breakpoints of each inversion. ***Scenario 2***: first in step 1, reads marked as unmapped after alignment are extracted, secondly in step 2, middle

2 Materials

2.1 Input Data

SRinversion detects short inversions based on NGS data with different read lengths from different sequencing platforms. And both paired end reads and single end reads are acceptable. The only requirement of the input NGS data is that the sequencing reads should be in FASTQ format [13].

Alignment results (*see* **Note 1**) in SAM/BAM format can also be processed by the framework, in which case the first step in the following section called *alignment* can be skipped (*see* **Note 2**).

2.2 Reference Genome

The default reference genome used by SRinversion for alignment of the sequencing reads is the Human reference Genome (hg19), which can be downloaded from the UCSC genome browser website [14].

2.3 Prerequirements

Before running the framework, (1) Perl 5 (in version 12 or above) need to be installed in the operating environment, (2) BWA (in version 0.7.5a-r405 or above) [15] should be installed and put in the directory called pub in the path where SRinversion located, and (3) Samtools (in version 0.1.19-44428cd or above) [16] need to be installed and put in the same directory as BWA.

2.4 SRinversion

The whole package of SRinversion can be downloaded from http://paed.hku.hk/genome/software/SRinversion/index.html.

3 Methods

3.1 Align Short Reads to Reference Genome

1. *Build index for reference sequences.* If the version of reference genome does not include index files needed for alignment using BWA, then index need to be built beforehand. Use the *index* package from BWA, and type command line '*bwa index ref.fa*', where *ref.fa* is the path of the corresponding reference genome file in FASTA format.
2. *Find the SA coordinates of the input reads.* Use the *aln* package from BWA. If the input data is single-end reads, then type command line '*bwa aln ref.fa input.fq > out.sai*', where *input.fq* is the path of the input single-end reads, and *out.sai* is the output file generated by BWA. If the input data is paired-end reads, then type command line '*bwa aln ref.fa input.1.fq > out.1.sai*' and command line '*bwa aln ref.fa input.2.fq > out.2.sai*', where *input.1.fq* and *input.2.fq* are paths of the pair of input reads, and *out.1.sai* and *out.2.sai* are paths of the corresponding output files.
3. *Generate alignments in SAM format.* If the input data is single-end reads, use the *samse* package from BWA, and type com-

mand line *'bwa samse –t n –r "readgroup" ref.fa out.sai input.fq > out.sam'*, where *out.sam* is the path of the output alignment file in SAM format. Here use parameter "*-t*" to set the number of threads to process this step. And use parameter "*-r*" to specify the name of the read groups for tracking.

4. *Change SAM format of the alignment files into BAM format.* Use Samtools to change the SAM format into corresponding binary format to reduce storage space. Type command line *'samtools view -Sb out.sam > out.bam'*.

3.2 Detect Short Inversions

SRinversion detect short inversions based on two scenarios. And the two scenarios should be processed separately.

1. *Detect short inversions in scenario 1.* Type command line *'perl SRinversion.pl -i out.bam -o sen1.chrID -s 1 -r ref.fa -c chrID'*, where *out.bam* is the path of the output alignment file from Subheading 3.1, *ref.fa* is the same reference genome used in alignment, and *sen1.chrID* is the path of the output file. Here parameter "*-s*" specify which scenario is considered for inversion identification. And in this step "*-s*" is set to 1. Parameter "*-c*" specify which chromosome to process (*see* **Note 3**).
2. *Detect short inversions in scenario 2.* Type command line *'perl SRinversion.pl -i out.bam -o sen2.chrID -s 2 -g 3 -r ref.fa -c chrID'* (*see* **Note 4**). Here parameter "-g" specify the gap length for sliding window to move on each time when splitting reads (*see* **Note 5**). To get the maximum efficiency, "-g" is recommended to set to 3.
3. *Merge results of the two scenarios.* Merge all results for the two scenarios in different chromosomes into one file for further filtering (*see* **Note 6**). To do this, type command *'cat sen1.chr* sen2.chr* > out.all'*.
4. *Filter inversions based on depth.* Basic depth filtering need to be performed to reduce false positives in the original results (*see* **Note 7**). To do this, type command *'awk '$6>=d' out.all > out.final'*, where *d* is the minimum number of supportive reads needed for an inversion event to be included (*see* **Note 8**).

4 Notes

1. Do not remove duplicated reads after alignment. Removing duplications are required for detection of SNVs and indels. However, SRinversion focuses on detecting signals in poorly mapped reads, which include reads that are marked as duplicates after alignment. So to retain as much information as possible, use original alignment results that include duplicated reads as input data for the framework.

2. Before running the framework, the paths of BWA and Samtools should be specified. There are two ways to do this: (1) put the original program files or links in directory called *pub* in the path where SRinversion is located in. (2) specify the absolute path of the two programs by setting parameters "*--bwa*" and "*--samtools*".
3. If parameter "*--chr*" is not set by users, then SRinversion will process all chromosomes (chromosome 1-22, and X, Y, M) one by one. To reduce time needed to run the program, specify each chromosome separately use "*--chr*" and process different chromosomes parallel.
4. Use parameter "*--clip*" to specify the minimum length of clipped ends to be included for analysis in scenario 1. In another words, if "*--clip*" is set to *n*bp, then in step 1 of the framework, only reads that are marked as *m*bp ($m > =n$) clipped after alignment will be extracted. For NGS reads with read length 100 bp, "--clip" is recommended to be set to 10, which is demonstrated to be the most suitable cutoff balancing sensitivity and specificity. To increase power in detecting smaller inversions, particularly ones shorter than 10 bp, "--clip" can be set smaller than 10 bp. However, it is notable that this might sacrifice specificities. Since in step 3, the split segments will be mapped to reference genome and segments smaller than 10 bp might result in multiple hits and bring in additional errors.
5. Use parameter "*--gap*" to specify gap length for sliding window to move on each time when splitting reads. The default value of "*--gap*" is 1 bp, which means removing sliding window along candidate read one base pair each time without any gaps. This will generate large amount of intermediate files since SRinversion generates one converted read and output both sequence and base quality of the read for each step of the sliding window. For instance, when using high coverage WGS reads with read length of 100 bp and depth of 30×, the estimated maximum storage space needed for one sample is around 500G. To reduce the storage, set a higher value for "*--gap*". It is clear that the higher "*--gap*" is set, the smaller amount of intermediate data will be generated by the framework. But introducing bigger gaps for sliding windows will decrease sensitivity of the program. To keep the sensitivity of the results acceptable, it is better to set the "*--gap*" smaller than 5 bp.
6. Set the parameter "*--minwin*" to specify minimum sliding window size in scenario 2. This works similar as parameter "*--gap*". When "*--minwin*" is set to 2 bp, the results achieve maximum sensitivity when all other conditions are the same. But this will produce a great deal of intermediate files and requires large

storage space for processing. To reduce the storage space needed for running the program, set a bigger "*--minwin*". But it is better to keep the value smaller than 10 bp to retain an acceptable sensitivity.

7. Depth filtering is required. And to avoid false positives caused by random sequencing errors in original NGS reads, the depth cutoff should be higher than 2 regardless of the coverage of original data. A cutoff of depth ≥5 is recommended for data with depth above 30×.

8. SRinversion can only estimate breakpoints of inversions at a resolution of around 5 bp. To increase the resolution to 1 bp and get the detailed sequence of the inversions, use Velvet [17] to assemble reads within the inversion regions into contigs, and map the contigs to the reference to figure out the precise breakpoints with the help of BLAST [18]. To do this, set the "--keep" parameter on to keep all the intermediate files. And reads for assembly can be extracted from the files in SAM format which stores all the candidate reads that are unmapped or with low mapping quality.

References

1. Lakich D, Kazazian HH Jr, Antonarakis SE et al (1993) Inversions disrupting the factor VIII gene are a common cause of severe haemophilia A. Nat Genet 5(3):236–241
2. Bondeson ML, Dahl N, Malmgren H et al (1995) Inversion of the IDS gene resulting from recombination with IDS-related sequences is a common cause of the Hunter syndrome. Hum Mol Genet 4(4):615–621
3. Feuk L, Carson AR, Scherer SW (2006) Structural variation in the human genome. Nat Rev Genet 7(2):85–97
4. Gimelli G, Pujana MA, Patricelli MG et al (2003) Genomic inversions of human chromosome 15q11-q13 in mothers of Angelman syndrome patients with class II (BP2/3) deletions. Hum Mol Genet 12(8):849–858
5. Osborne LR, Li M, Pober B et al (2001) A 1.5 million-base pair inversion polymorphism in families with Williams–Beuren syndrome. Nat Genet 29(3):321–325
6. Chen K, Wallis JW, McLellan MD et al (2009) BreakDancer: an algorithm for high-resolution mapping of genomic structural variation. Nat Methods 6(9):677–681
7. He F, Li Y, Tang YH et al (2016) Identifying micro-inversions using high-throughput sequencing reads. BMC Genomics 17(Suppl 1):4
8. Rausch T, Zichner T, Schlattl A et al (2012) DELLY: structural variant discovery by integrated paired-end and split-read analysis. Bioinformatics 28(18):i333–i339
9. Trappe K, Emde AK, Ehrlich HC et al (2014) Gustaf: detecting and correctly classifying SVs in the NGS twilight zone. Bioinformatics 30(24):3484–3490
10. Ye K, Schulz MH, Long Q et al (2009) Pindel: a pattern growth approach to detect break points of large deletions and medium sized insertions from paired-end short reads. Bioinformatics 25(21):2865–2871
11. Stankiewicz P, Lupski JR (2010) Structural variation in the human genome and its role in disease. Annu Rev Med 61:437–455
12. Chen R, Lau YL, Zhang Y et al (2016) SRinversion: a tool for detecting short inversions by splitting and re-aligning poorly mapped and unmapped sequencing reads. Bioinformatics 32(23):3559–3565
13. Cock PJA, Fields CJ, Goto N et al (2009) The Sanger FASTQ file format for sequences with quality scores, and the Solexa/Illumina FASTQ variants. Nucleic Acids Res 38(6):1767–1771
14. Kent WJ, Sugnet CW, Furey TS et al (2002) The human genome browser at UCSC. Genome Res 12(6):996–1006

15. Li H, Durbin R (2009) Fast and accurate short read alignment with Burrows–Wheeler transform. Bioinformatics 15:1754–1760
16. Li H, Handsaker B, Wysoker A et al (2009) The sequence alignment/map (SAM) format and SAMtools. Bioinformatics 25:2078–2079
17. Daniel RZ, Birney E (2008) Velvet: algorithms for de novo short read assembly using de Bruijn graphs. Genome Res 18:821–829
18. Boratyn GM, Camacho C, Cooper PS et al (2013) BLAST: a more efficient report with usability improvements. Nucleic Acids Res 41:W29–W33

Chapter 9

Detection of CNVs in NGS Data Using VS-CNV

Nathan Fortier, Gabe Rudy, and Andreas Scherer

Abstract

Copy number variations have been linked to numerous genetic diseases including cancer, Parkinson's disease, pancreatitis, and lupus. While current best practices for CNV detection often require using microarrays for detecting large CNVs or multiplex ligation-dependent probe amplification (MLPA) for gene-sized CNVs, new methods have been developed with the goal of replacing both of these specialized assays with bioinformatic analysis applied to next-generation sequencing (NGS) data. Because NGS is already used by clinical labs to detect small coding variants, this approach reduces associated costs, resources, and analysis time. This chapter provides an overview of the various approaches to CNV detection via NGS data, and examines VS-CNV, a commercial tool developed by Golden Helix, which provides robust CNV calling capabilities for both gene panel and exome data.

Key words Copy number variation, Next generation sequencing

1 Introduction

Prior to the advent of genome-wide analyses circa 2004, it was not widely recognized that copy number variation (CNV) was both widespread throughout the human genome, and responsible for human genetic variation and disease. The first time that a CNV was confirmed to be causative for a genetic disease was back in the early 1990s, when it was reported that the hereditary neuropathy Charcot–Marie–Tooth disease was associated not with a flawed version of a particular gene, but instead was due to the presence of an extra copy of the normal version of PMP22 contained within a duplication of a large region of chromosome 17 [1]. Since then our understanding CNVs and their impact in disease has grown considerably, with analyses revealing over 10,000 CNVs overlapping over 1000 genes [2, 3]. In 2007, Lupsky observed that locus-specific mutation rates for genomic rearrangements appear to be two to four orders of magnitude greater than nucleotide-specific rates for base substitutions [4].

Derek M. Bickhart (ed.), *Copy Number Variants: Methods and Protocols*, Methods in Molecular Biology, vol. 1833, https://doi.org/10.1007/978-1-4939-8666-8_9, © Springer Science+Business Media, LLC, part of Springer Nature 2018

CNVs are associated with a broad range of pathological conditions and complex traits, which are manifest in nearly all organs and tissues. Everything from Parkinson's disease, pancreatitis, lupus, and even susceptibility to HIV infection are associated with CNVs [5]. As more and more disorders and diseases associated with CNVs are identified, and as NextGen sequencing is more widely utilized in clinical practice, tools that will allow rapid, sensitive, cost effective, genome-wide identification and reporting of CNVs are now needed.

It is now recognized that, in terms of the number of nucleotides, CNVs account for more differences between human genomes than the more extensively studied single-nucleotide differences. A 2010 study estimated that CNVs may account for 13% of the human genome [5]. Currently, arrays for comparative genomic hybridization and SNP arrays are most commonly used in structural variation studies. However, the information that arrays can reveal about structural variation is limited because they can detect only sequences that match the oligonucleotide probes used to make them, and these probes are usually biased against "difficult," highly repetitive regions. Custom arrays with tens of millions of probes may find variants as small as 500 bases, but the use of such arrays is economically not feasible for studies that require large sample sizes. For the most commonly used arrays, the size limit of detection is usually much higher, generally on the order of 5 kb, and even greater for highly repetitive sequences. In addition to these limitations, while arrays can detect that a sample has more or fewer copies of a region compared to a reference genome, they generally cannot determine an absolute number. However, the availability of next-generation technologies to clinicians and researchers offers a potential path to circumvent these limitations. With the right software tools, clinicians and researchers will be able to begin mining short-read sequencing data for structural variation. While there are a wide range of different approaches to CNV detection, many of the algorithms share similarities and use common strategies to solve the various related sub-problems. Generally, CNV detection methods incorporate three major steps. First, data preprocessing is performed to correct for biases in the data and create a baseline for detecting variation. Second and third, algorithms need to assign copy-number states and define the boundaries of multitarget events using a segmentation algorithm. Different methods apply these steps in different orders.

Golden Helix, Inc., (GHI) has developed VS-CNV, which allows clinicians and researchers to detect CNVs based on NextGen sequencing (NGS) data. Today, this type of data is already widely used to analyze single nucleotide variations (SNVs). With VS-CNV, one can conduct analysis of SNVs and CNVs on a single data set. This streamlines the analysis workflow since clinicians no longer need to separately conduct the CNV analysis using micro arrays, thus eliminating the cost associated with additional procedures. Finally, VS-CNV increases the sensitivity of CNV analysis, allowing users to detect much smaller CNV events than would otherwise be detectable using arrays. This chapter provides an overview of the problem of CNV detection on NGS data, examines various approaches to problem, and presents VS-CNV, a commercial tool developed by Golden Helix, which allows clinicians and researchers to reliably detect CNVs from NGS data on both gene panels and exomes. In a third-party evaluation, VS-CNV was shown to have 100% concordance with MLPA for the detection of exon-level CNVs in LDLR, and, in additional experiments performed by the authors, VS-CNV was shown to have high sensitivity and precision for both gene panels and full exomes. VS-CNV's robust normalization procedure allows it to achieve high sensitivity, even on highly mutated cancer samples, on which other algorithms are not useful. Additionally, VS-CNV is capable of calling small single exon events with high precision. These smaller CNV events are often ignored in most testing scenarios, since they are not detected by existing microarray paradigms, or they are called on a limited number of genes using MLPA. The advancement of precision medicine requires the adoption of genetic tests that are economical and comprehensive in their detection of relevant genomic mutations. The use of VS-CNV can help achieve these goals by detecting CNVs as part of existing genomic tests based on NGS gene panels and whole-exome sequencing.

2 Materials

2.1 Data Preprocessing

Data preprocessing involves correcting the data for systematic biases and normalizing it to create a baseline for detecting variation. The two most common methods for addressing systematic bias are GC-content and mappability correction. The most common methods for normalization are principle component analysis (PCA), and reference sample normalization.

One source of bias in the coverage data is CG-content bias. It is known that regions with high or low GC-content tend to have lower mean read depth due to PCR efficiency in amplification. When correcting for GC-bias CNV calling algorithms generally will either filter out regions with extreme GC-content or perform normalization to account for the bias. Algorithms that use the filtering approach include XHMM and OncoSNP-SEQ [6, 7], while algorithms using normalization to account for GC content include CLAMMS, ReadDepth, Patchwork, and Control-FREEC [8–11]. In the CLAMMS algorithm, normalization is performed by dividing the coverage at a region by the median coverage of regions with similar GC-content. While this kind of normalization effectively corrects for GC-bias, it does cause the algorithm to incur an additional computational expense when compared to a simple filtering approach.

Another source of bias in the coverage data is mappability bias. Mappability for a given region is the probability that a read originating from the region is unambiguously mapped to it. Regions with low mappability tend to produce more ambiguous reads, which can cause errors in CNV detection. Generally, algorithms will address mappability bias by filtering out low mappability regions. Methods that address mappability bias in this way include as CODEX, Control-FREEC, and OncoSNP-SEQ [7, 8, 12].

Several CNV detection algorithms perform their primary normalization via principle component analysis (PCA) on the coverage data. PCA uses an orthogonal transformation to convert a set of observations into a set of linearly uncorrelated variables called principle components. The CONIFER and XHMM algorithms perform normalization using PCA by removing the *k* strongest principle components [6, 13]. As an alternate to PCA, it is also possible to perform normalization using a set of reference samples. This is done by using deviation from the average coverage in the reference samples as an indicator of CNV occurrence. Generally, this is done by computing evidence metrics, such as a Z-score, relative to the control samples. This approach normalizes out biases present across the reference samples, thereby reducing or eliminating the need to explicitly correct for systematic biases such as GC-content and mappability. Algorithms that rely on reference samples for CNV detection include CoNVaDING, VisCap, CLAMMS, and CNVkit [11, 14–16].

While PCA based normalization has the advantage of handling varied and even unknown sources of noise in the data, the approach has two major disadvantages compared to reference sample normalization in a clinical setting. First, it requires significantly more samples to provide robust results (at least 50 samples according to XHMMs documentation). Clinical labs may have as few as 15–20 samples as they validate and configure a test. Reference sample-based normalization can provide reasonable results with far fewer samples. Secondly, the choosing of the *k* strongest principle components to factor out of the data is a somewhat subjective parameter, yet highly influential on the final result. For clinical validation of bioinformatics methods in a genetic test, algorithms should be robust. This means that small changes in inputs will not lead to dramatically different results. Additionally, they need to be as transparent as possible in regards of the usage of intermediate values and metrics. Each false negative (known CNV missed by the algorithm) must be investigated and understood to characterize the limits of a test. For these reasons, the black-box nature of PCA may be inferior to the reference sample normalization approach for an algorithm in the clinical context.

2.2 State Assignment

During the state assignment step, a copy number state is assigned to each target (or each segment, if segmentation is performed first). The classification problem requires some empirical criteria for assigning a copy number state to a given region. Some algorithms rely on empirically defined thresholds to determine copy number state, while others use Hidden Markov Models (HMM) for classification. Thresholding is the simplest method for CNV classification. This approach involves setting thresholds for one or more of the metrics and calling a CNV if the metrics at the target fall above or below the thresholds. Thresholding is used by CoNVaDING, ReadDepth, Patchwork, Control_FREEC, and BIC-Seq [8–10, 14, 17]. Alternatively, classification can be performed using hidden Markov models (HMM). HMMs are statistical models that represent the system as a Markov process with hidden states [18]. In an HMM, the state of the system is not directly observable but a single evidence variable, which is dependent on the state, can be observed. Algorithms that use HMMs for classification include XHMM, CANOES, and CLAMMS [6, 11, 19].

When compared to thresholding approaches for state assignment, HMMs have several key advantages. First, HMMs account for conditional dependencies between the states of adjacent target regions, increasing the probability that a target will have the same state as its neighbor. Second, the probabilistic nature of HMMs allows us to quantify the uncertainty of CNV calls by assigning a probability to each called region. However, unlike thresholding methods, HMMs cannot easily incorporate multiple evidence variables. This shortcoming has led many researchers to rely on thresholding methods despite HMM's other advantages. Consequently, VS-CNV relies on a general probabilistic model capable of incorporating multiple evidence metrics, allowing it to combine the advantages of thresholding and HMM methods.

2.3 Large Event Calling

Large events must be constructed by merging targets in the same contiguous region into a single event with well-defined boundaries. The most common approach is a simple merging procedure of consecutive targets with the same copy number state. This approach is used by CLAMMS, CANOES, CoNVaDING, OncoSNP-SEQ, and XHMM [6, 7, 11, 14, 19]. Unfortunately, these methods fail to reliably call large CNVs as one contiguous event, which is desirable from a clinical interpretation perspective. Instead, these methods will call larger CNV events as a group of smaller events separated by outliers.

Other methods, especially those focused on calling large events, up to and including chromosome level aneuploidy, perform segmentation before calling copy number state. The most common segmentation algorithm is Circular Binary Segmentation (CBS). CBS performs segmentation by iteratively computing segments to maximize the variance between segments while minimizing the variance within each segment [20]. It is used by ExomeCNV, VarScan2, and Patchwork [9, 21, 22]. Unfortunately, algorithms performing segmentation before event classification are only capable of detecting large events spanning multiple targets and are unable to detect critical single exon CNVs known to occur in clinical samples.

3 Methods

VS-CNV integrates multiple metrics in order to determine, if a CNV event is present. These metrics include:

- *Z-score*: The Z-score measures the number of standard deviations a target is from the reference sample mean. It is computed by subtracting the normalized read depth of the reference samples from the normalized depth for the sample of interest and dividing the result by the standard deviation of the reference samples. A high Z-score is indicative of a duplication event, while a lower Z-score is evidence for a deletion event. The Z-scores are also used to compute *p*-values for each called event. The *p*-value for an event measures the probability of Z-scores at least as extreme assuming the event targets are diploid, and can be useful for evaluating call quality.
- *Ratio*: The ratio is computed for a given target by dividing the normalized read depth for the sample of interest by the normalized mean depth over the reference samples. If no CNV event is present, the sample of interest should have the same normalized depth as the reference samples, indicating a ratio value close to 1, while homozygous deletions, heterozygous deletions, and duplications will have ratio values around 0, 0.5, and 1.5, respectively. Unlike the Z-score, the ratio gives us the ability to differentiate between homozygous, and heterozygous deletion events.
- *Variant allele frequency (VAF)*: VAF is the allelic fraction of sequence reads at the SNP locus for the allele that differs from the reference sequence. In a normal genome there are four genotype possibilities at a given locus: AA, AB, BA, and BB with VAF values of 0, 0.5, 0.5, and 1, respectively. However, when the copy number deviates from 2, the VAF has other possible values, which depend on the copy number.

The first two metrics are computed from normalized coverage and provide the primary evidence used to identify CNV events.

The combination of the Z-score and ratio allows us to detect CNV events ranging from small single exon events to large whole

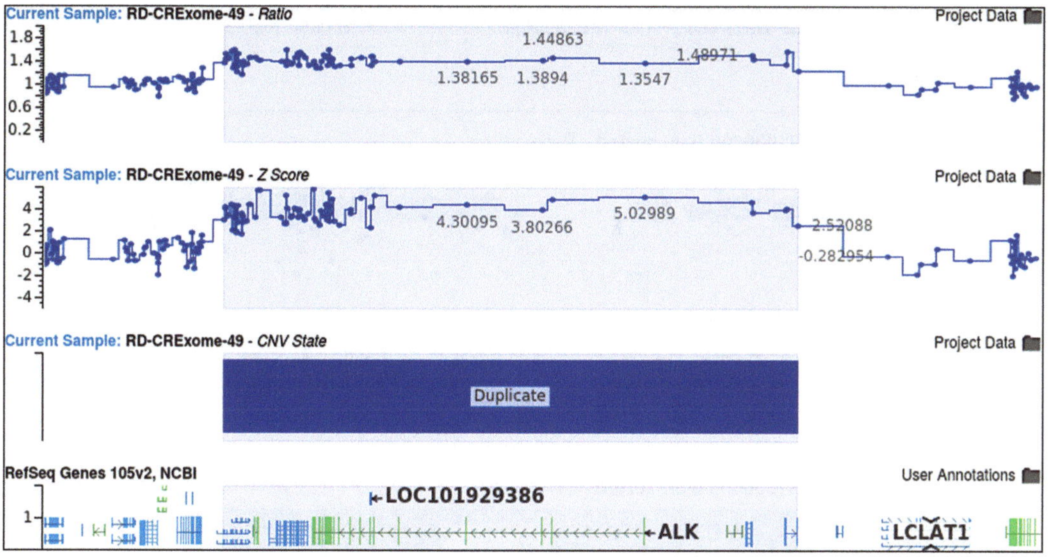

Fig. 1 Example of a large multigene duplication event of the ALK gene

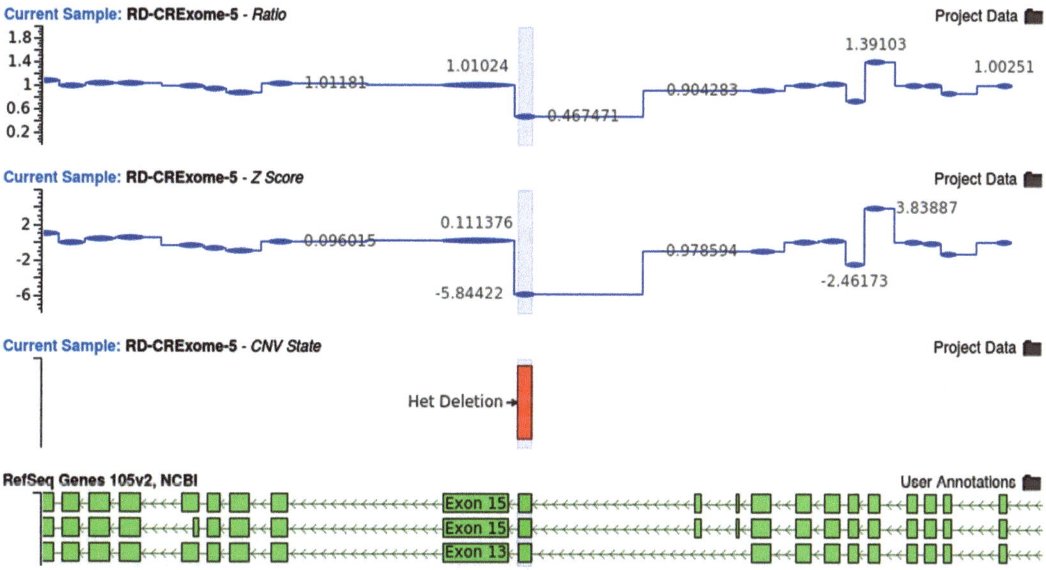

Fig. 2 A detected single exon deletion of the FHOD1 gene. The −6 Z-score is an indicator of the confidence that this deletion truly exists in the data

chromosome events. Fig 1 shows a large multigene duplication event, encompassing the ALK gene. The large Z-score indicates that targets within this event are around 5 standard deviations from the reference samples. These large Z-scores, combined with the ratio values centering around 1.5, provide strong evidence for this duplication. Figure 2 shows a heterozygous deletion of a single exon in the gene FHOD1. With a Z-score nearly 6 standard deviations from

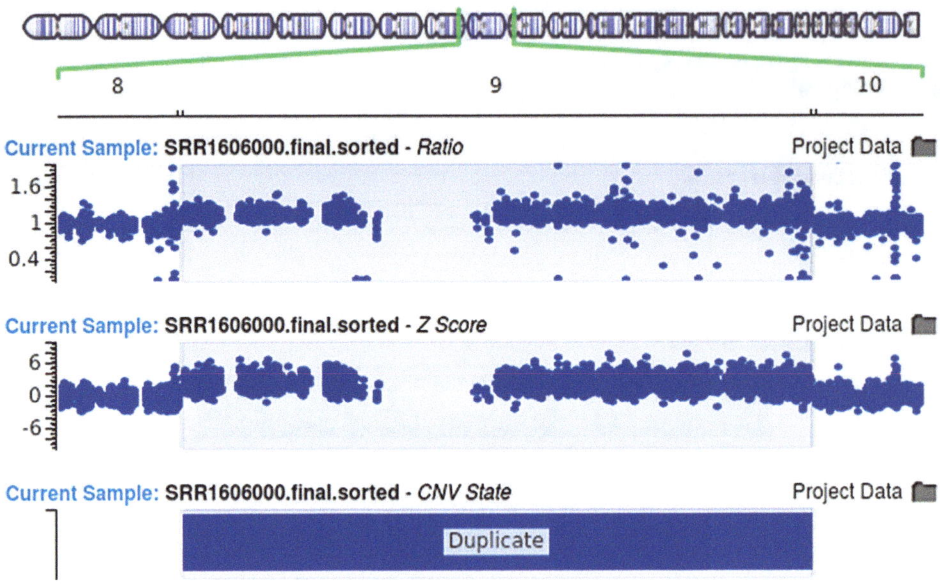

Fig. 3 Detection of a whole chromosome duplication on chromosome 9

the reference samples and a ratio very close to the 0.5 value expected for heterozygous deletions, we have excellent evidence for this single exon event. Figure 3 shows a duplication of chromosome 9. This whole chromosome duplication is supported by an elevated Z-score and ratio spanning the entire chromosome. In comparison, you see in Fig. 4 a textbook call for a deletion.

While the Z-score and ratio provide the primary evidence for CNV calls, the VAF can also provide important information, both during the normalization process, and when verifying called CNVs. The VAF has two important uses in our approach. First, regions with abnormal VAF are be excluded from the normalization process, which helps prevent skewing of the normalized read depth due to large chromosomal events. Second, it can provide supporting evidence used to confirm true events, and reduce false positive calls. For example, deletion events should have bimodal distributions, with peaks around 0 and 1, while triploid duplications will have a multimodal distribution, with VAFs centered around 0, 1/3, 2/3, and 1.

3.1 Third Party Benchmark

An independent evaluation of VS-CNV was performed by Iacocca et al., who analyzed 388 samples from patients with familial hypercholesterolemia (FH), a heritable disorder linked to autosomal codominant mutations in the LDL receptor gene (LDRL) [23]. Current best practices for diagnosing FH include targeted NGS panels for detection of small variants in conjunction with MLPA in LDLR for the detection of CNVs. The goal of the study was to evaluate the potential of replacing MLPA with VS-CNV, thereby eliminating the need for an additional assay, and greatly reducing

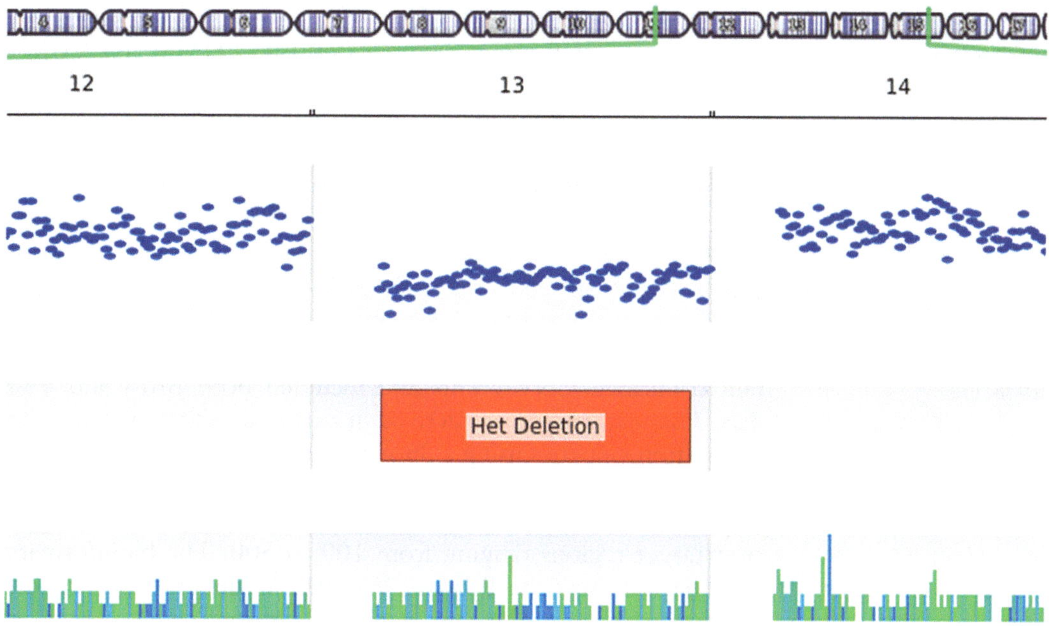

Fig. 4 An example heterozygous deletion call

the cost of analysis. The authors reported 100% concordance for the detection of exon-level CNVs in LDLR between VS-CNV and MLPA.

The study was conducted with the following approach. DNA was isolated from blood samples of the 388 individuals using the Puregene DNF Blood Kit and was sequenced for 73 genes, including all major and minor genes associated with FH. Whole-exon deletion and duplication events in LDLR were detected by the orthogonal methods of MLPA and VS-CNV. VS-CNV detected CNVs in LDLR for 38 of the 388 patients. These CNVs were found to be in 100% concordance with MLPA. Additionally, all samples testing negative for CNVs according to MLPA were also negative according to VS-CNV. Thus, VS-CNV resulted in 100% diagnostic sensitivity and specificity, using MLPA as the "gold standard" reference method.

Based on these results, the authors suggest that MLPA can be replaced by VS-CNV in the diagnostic workup for FH, eliminating the cost of this expensive additional assay. Specifically, in the case of this study, eliminating MLPA would reduce the diagnostic cost by around $80 per patient, for a total cost reduction of $31,000. Additionally, the *use of VS-CNV allowed the authors to extend CNV analysis to all FH-associated genes at no additional cost.* In contrast, it is not economically feasible to apply MLPA to genes outside of LDLR, as this would require additional assays for each analyzed gene.

3.2 Comparisons of Precision

VS-CNV was compared to four alternative approaches to CNV detection on NGS target data: CONIFER, XHMM, CLAMMS, and CoNVaDING. The sensitivity and precision of these algorithms was evaluated on two datasets. The first consisted of 63 samples sequenced on a hereditary cancer gene panel, and the second consisted of 202 exome samples, of which 109 are highly mutated myeloma samples. VS-CNV resulted in superior sensitivity for both datasets, while maintaining competitive precision (see below for details).

3.3 Comparison Datasets

For these comparisons, two datasets were used, both of which contain confirmed CNV events and included both BAM and VCF files. The first dataset was derived in a clinical validation setting by Prevention Genetics using a 35-gene panel used to discern hereditary risk of cancer. The sequencing was performed with an Illumina sequencer with 100 bp paired-end reads with sample mean depths over target regions ranging from 100 to 500. The bioinformatic pipeline used was the Burrows-Wheeler Alignment Tool in conjunction with the GATK variant caller. It consists of 63 samples, of which 38 contain confirmed CNV events based on MLPA and high density microarray assays. Of these samples, 14 contain single exon CNVs, allowing us to evaluate the ability of the algorithms to detect small single target events that are required to be detected by the clinical test.

The second dataset was obtained from the International Cancer Genome Consortium (ICGC) [24]. This data was curated as part of a longitudinal observation study of myeloma patients performed by the Multiple Myeloma Research Foundation [25]. It consists of BAM and VCF files for 202 samples, 109 of which are highly mutated myeloma samples containing confirmed CNV events. These samples were sequenced using the Illumina TruSeq Exome Enrichment Kit. Most of these samples contain large whole-chromosome events, and for many of the samples, most of the exome is in a deletion or duplication state. Consequently, this dataset is particularly challenging for traditional approaches to CNV detection. For these samples, the truth set is based on CNVs called on whole-genome sequencing of the same set of samples. Therefore, this dataset represents the opposite end of the CNV size spectrum and only contains very large events detected by the WGS CNV caller.

3.4 Comparison Results

Tables 1 and 2 show the performance of the algorithms on the cancer panel and myeloma exome datasets, respectively (*see* **Note 1**). For the cancer panel dataset, XHMM achieved superior sensitivity and precision when compared to CoNIFER, CLAMMS, and CoNVaDING, but failed to call any of the single exon events. While CoNVaDING called a large percentage of the single exon events, it produced many false positive calls, resulting in low precision.

Table 1
Performance on prevention cancer panel

Cancer panel	Sensitivity (%)	Precision (%)	Single exon events called (%)
VS-CNV	97.6	89.7	42.9
CoNIFER	83.7	0	15.4
XHMM	92.6	88.9	0
CLAMMS	70.8	72.2	15.4
CoNVaDING	72.3	29.7	84.6

Table 2
Performance on myeloma exomes

Myeloma exomes	Sensitivity (%)	Precision (%)	Runtime
VS-CNV	65.9	93.7	Hours
CoNIFER	0.3	64.1	Minutes
XHMM	4.9	74.0	Minutes
CLAMMS	18.1	98.8	Minutes
CoNVaDING	0.7	88.5	Weeks

In contrast, VS-CNV managed to outperform all others in terms of sensitivity and precision, while still calling nearly half of the single exon events.

For the myeloma exomes, VS-CNV achieved the highest sensitivity, while CLAMMS achieved the highest precision. In terms of runtime, XHMM had the best performance, finishing within minutes, while CoNVaDING had the worst, taking nearly 2 weeks to complete. Our algorithm's runtime was slower than that of CLAMMS, CoNIFER, and XHMM, but was faster than CoNVaDING.

3.5 Comparison Discussion

These results demonstrate the superiority of VS-CNV, when compared to existing CNV detection methods. For the cancer panel data, VS-CNV outperformed all competing algorithms in terms of both sensitivity and precision, while calling a large percentage of single exon events. In contrast, no competing approach was capable of calling a significant percentage of single exon events, without also calling a large number of false positives. This indicates that VS-CNV is the only method capable of providing clinical-grade, high precision detection of both large and small CNV events on gene panel data.

The results for the myeloma exome samples are even more striking. Because the samples in this dataset are highly mutated tumors, no competing approach was able to achieve a sensitivity higher than 20%. Other authors are aware of this issue, and the CLAMMS algorithm even issues a disclaimer, warning users to not run the algorithm on cancer samples. Such a low true-positive rate makes these algorithms unusable for detecting CNVs present in highly mutated cancer samples. In contrast, VS-CNV achieved a sensitivity *over three times higher* than the highest performing competitor, while maintaining excellent precision. In terms of runtime, VS-CNV required less than a day to call CNVs in over one hundred full exome samples. This is less time than that is required to run the initial coverage computations, which are required by all algorithms. Thus, in the context of tumor exomes, VS-CNV is the only method likely to provide significant utility in a clinical and research setting.

4 Notes

1. Performance of the algorithms on both datasets was evaluated in terms of sensitivity and precision at the base pair level. For the set of CNV calls produced by each algorithm, each base pair was classified as either true positive (TP), true negative (TN), false positive (FP), or false negative (FN). The number of base pairs falling into these categories was then used to compute the sensitivity and precision for each algorithm as follows:

$$\text{Sensitivity} = \frac{\text{TP}}{\text{TP} + \text{FN}} \quad \text{Precision} = \frac{\text{TP}}{\text{TP} + \text{FP}}$$

For the cancer panel data, the percentage of single exon events successfully called by each of the algorithms was also computed. For the myeloma exomes, the algorithms were also compared in terms of runtime, with each algorithm's runtime classified as minutes, hours, days, or weeks.

References

1. Chance PF, Pleasure D (1993) Charcot-Marie-tooth syndrome. Arch Neurol 50:1180–1184
2. Conrad DF, Pinto D, Redon R et al (2010) Origins and functional impact of copy number variation in the human genome. Nature 464:704
3. Redon R, Ishikawa S, Fitch KR et al (2006) Global variation in copy number in the human genome. Nature 444:444–454
4. Lupski JR (2007) Genomic rearrangements and sporadic disease. Nat Genet 39:S43–S47
5. Stankiewicz P, Lupski JR (2010) Structural variation in the human genome and its role in disease. Annu Rev Med 61:437–455
6. Fromer M, Moran JL, Chambert K et al (2012) Discovery and statistical genotyping of copy-number variation from whole-exome sequencing depth. Am J Hum Genet 91:597–607
7. Yau C (2013) OncoSNP-SEQ: a statistical approach for the identification of somatic copy number alterations from next-generation

sequencing of cancer genomes. Bioinformatics 29(19):2482–2484

8. Boeva V, Popova T, Bleakley K et al (2012) Control-FREEC: a tool for assessing copy number; allelic content using next-generation sequencing data. Bioinformatics 28(3): 423–425
9. Mayrhofer M, DiLorenzo S, Isaksson A (2013) Patchwork: allele-specific copy number analysis of whole-genome sequenced tumor tissue. Genome Biol 14(3):1
10. Miller CA, Hampton O, Coarfa C et al (2011) ReadDepth: a parallel R package for detecting copy number alterations from short sequencing reads. PLoS One 6(1):e16327
11. Packer JS, Maxwell EK, O'Dushlaine C et al (2016) CLAMMS: a scalable algorithm for calling common and rare copy number variants from exome sequencing data. Bioinformatics 32(1):133–135
12. Jiang Y, Oldridge DA, Diskin SJ et al (2015) CODEX: a normalization and copy number variation detection method for whole exome sequencing. Nucleic Acids Res 43(6):e39
13. Krumm N, Sudmant PH, Ko A et al (2012) Copy number variation detection and genotyping from exome sequence data. Genome Res 22(8):1525–1532
14. Johansson LF, Dijk F, Boer EN et al (2016) CoNVaDING: single exon variation detection in targeted NGS data. Hum Mutat 37(5): 457–464
15. Pugh TJ, Amr SS, Bowser MJ et al (2015) VisCap: inference and visualization of germ-line copy-number variants from targeted clinical sequencing data. Genet Med 18(7):712–719
16. Talevish E, Shain AH, Botton T et al (2016). CNVkit: Genome-Wide copy number detection and visualization from targeted DNA sequencing. PLoS Computational Biology 12(4):e1004873
17. Xi R, Luquette J, Hadjipanayis A et al (2010) BIC-seq: a fast algorithm for detection of copy number alterations based on high-throughput sequencing data. Genome Biol 11(1):1
18. Koller D, Friedman N (2009) Probabilistic graphical models: principles and techniques. MIT Press, Cambridge
19. Backenroth D, Homsy J, Murillo LR et al (2014) CANOES: detecting rare copy number variants from whole exome sequencing data. Nucleic Acids Res 42(12):e97
20. Olshen AB, Venkatraman E, Lucito R et al (2004) Circular binary segmentation for the analysis of array-based DNA copy number data. Biostatistics 5(4):557–572
21. Koboldt DC, Zhang Q, Larson DE et al (2012) VarScan 2: somatic mutation and copy number alteration discovery in cancer by exome sequencing. Genome Res 22(3):568–576
22. Sathirapongsasuti JF, Lee H, Horst BA et al (2011) Exome sequencing-based copy-number variation and loss of heterozygosity detection: ExomeCNV. Bioinformatics 27(19):2648–2654
23. Iacocca MA, Wang J, Dron JS et al (2017) Use of next-generation sequencing to detect LDLR gene copy number variation in familial hypercholesterolemia. J Lipid Res 58(11):2202–2209
24. Zhang J, Baran J, Cros A et al (2011) International cancer genome consortium data portal—a one-stop shop for cancer genomics data. Database (Oxford). https://doi.org/10.1093/database/bar026
25. Craig DW, Liang W, Venkata Y et al (2013) Interim analysis of the Mmrf Commpass trial, a longitudinal study in multiple myeloma relating clinical outcomes to genomic and immunophenotypic profiles. Blood 122(21):532

Chapter 10

Structural Variant Breakpoint Detection with novoBreak

Zechen Chong and Ken Chen

Abstract

Structural variations (SVs) are an important type of genomic variants and always play a critical role for cancer development and progression. In the cancer genomics era, detecting structural variations from short sequencing data is still challenging. We developed a novel algorithm, novoBreak (Chong et al. Nat Methods 14:65–67, 2017), which achieved the highest balanced accuracy (mean of sensitivity and precision) in the ICGC-TCGA DREAM 8.5 Somatic Mutation Calling Challenge. Here we describe detailed instructions of applying novoBreak (https://github.com/czc/nb_distribution), an open-source software, for somatic SVs detection. We also briefly introduce how to detect germline SVs using novoBreak pipeline and how to use the Workflow (https://cgc.sbgenomics.com/public/apps#ZCHONG/novobreak-commit/novobreak-analysis/) of novoBreak on the Seven Bridges Cancer Genomics Cloud.

Key words Structural variations, Algorithm, Next generation sequencing data analysis, DNA sequence analysis, Genomic rearrangement, De novo assembly, *k*-mer, Genetic variation

1 Introduction

Structural variations (SVs) are an important type of genomic variants. De novo SVs are major contributors for genome evolution and a wide array of diseases [1]. SVs are very common in different types of cancers [2–4]. Somatic SVs can be recurrent in a high frequency [5], which is ideal for drug target. For example, the drug imatinib can specifically target BCR-ABL1 gene fusion in Chronic Myelogenous Leukemia (CML) patients.

The advent of high-throughput next generation sequencing technologies enables detecting all types of variants including SVs at single-base pair resolution. As a result, an unprecedented landscape of SVs has been discovered in both healthy and unhealthy genomes. However, current sequencing-based computational methods [6–12] are limited in sensitivity and comprehensiveness [13] due to several reasons: (1) there are many types of SVs: deletions, duplications, insertions, inversions, translocations, etc., each having a distinctive rearrangement pattern when the sequenced reads are compared to the reference genome [14]; (2) the sizes of SVs vary from tens of

Derek M. Bickhart (ed.), *Copy Number Variants: Methods and Protocols*, Methods in Molecular Biology, vol. 1833,
https://doi.org/10.1007/978-1-4939-8666-8_10,

base pairs to several mega base pairs, thus detection requires genome-wide analysis of all the sequencing reads; (3) complex SVs such as chromothripsis [15] and chromoplexy [16] may result in complex rearrangement architecture that involve multiple chromosomes, breakpoints, and clusters of mutations near breakpoints; (4) current NGS reads are relatively short and difficult to map accurately to the reference genome, especially when containing breakpoints; (5) the impurity or intratumor heterogeneity of tumor tissues reduce the coverage of the variant alleles (particularly for heterozygous or subclonal SVs).

One major approach for SV detection is resequencing based, i.e., to align reads to a reference genome and then identify signals in discordant read pairs [6, 7], read depths [8], split reads [9], or their combinations [11]. Another approach is through local assembly of aligned and partially aligned reads in candidate SV regions discovered a priori [10, 12]. These methods depend heavily on the quality of short read alignment, which are often limited for reads spanning breakpoints and substantially different from the reference genome. Theoretically, whole genome de novo assembly approaches [17] do not rely on reference-alignment and could be less biased. However, assembling the whole genome is computationally challenging [18] and the results are limited by repeats, heterozygotes, polyploidy, read length, and sequencing coverage.

We developed a novel method, novoBreak [19], which directly identifies breakpoints from clusters of reads that share a set of *k*-mers (contiguous nucleotide sequences of length *k*) uniquely present in a subject genome (e.g., a tumor genome) but not in the human reference genome or any control data (e.g., a matched normal genome) (Fig. 1). When applied to somatic breakpoint detection from a pair of matched tumor and normal genomes, novoBreak first constructs a hash table from the tumor reads, containing all the *k*-mers, their host reads and frequencies in the set. Next, it filters out *k*-mers representing reference alleles or sequencing errors, and retains those representing variants or novel sequences not present in the reference genome (Fig. 1a). It then queries the normal reads and further categorizes the *k*-mers into two classes: (1) germline, those present in both the tumor and the normal genome, and (2) somatic, those present in the tumor but not the normal genome. After that, novoBreak identifies clusters of read pairs, each of which spans a somatic breakpoint, and assembles each cluster into contigs (Fig. 1b). By comparing the resulting high-quality contigs with the reference, novoBreak identifies breakpoints and characterize associated SVs (Fig. 1c). Finally, novoBreak quantifies the amount of the evidence at each breakpoint and outputs a final report (Fig. 1d).

If there is no matched normal genome and you would like to detect SVs from a single sample, you can simulate a mocked matched normal genome. This strategy can be used to identify germline SVs from a single genome.

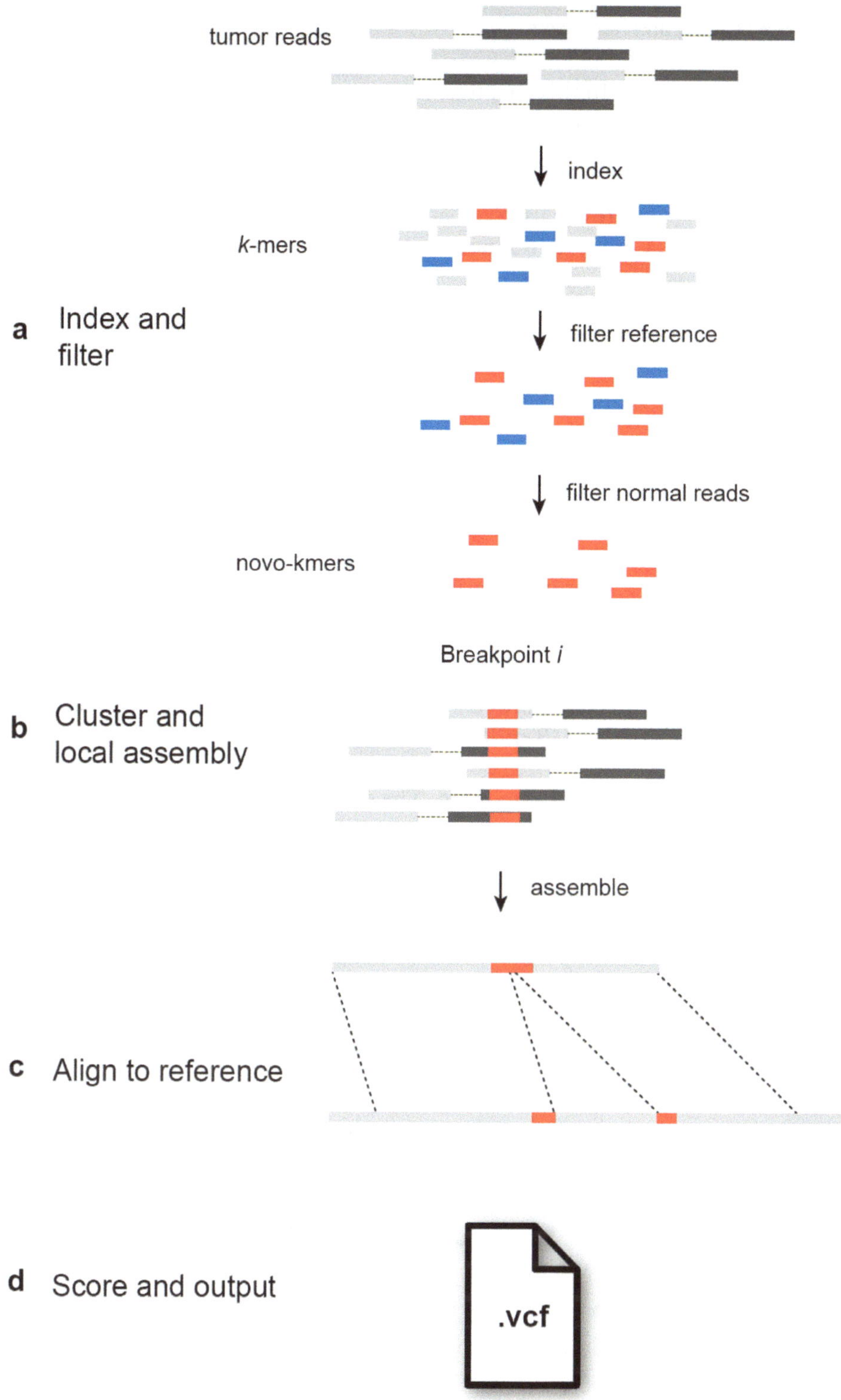

Fig. 1 The workflow begins by (**a**) indexing and identifying all of the unique kmers present in a read dataset but not in a reference genome. The reads that are unique to the dataset (**b**) are then clustered by kmers and assembled into contigs. The contigs (**c**) are then aligned to the reference genome to identify the source of structural variation and the location of the novel sequence insertion. Finally (**d**) the identified variants are scored and output in the VCF file format

In this chapter, we present a step-by-step protocol for identifying somatic SVs from tumor–normal matched studies. We will also introduce a simply modified version of the somatic pipeline for detecting germline SVs from a single sample. Finally, we demonstrate the usage of novoBreak on a cloud platform—Seven Bridges Cancer Genomics Cloud.

2 Materials

We describe the equipment and equipment setup in this section.

2.1 Equipment

1. Data. The protocol can be applied to Illumina paired-end sequencing data, either whole genome sequencing or whole exome sequencing data. The inputs of novoBreak include two BAM files (*see* **Note 1**) and a Reference sequence (*see* **Note 2**) that used for mapping to generate the BAM files.
2. NovoBreak software (https://github.com/czc/nb_distribution).
3. Git (https://git-scm.com/downloads).
4. Hardware (64-bit computer running Linux Operating System; 64 GB of RAM (100 GB preferred); 1 TB disk storage).

2.2 Equipment Setup

1. Hardware setup. The software used in this protocol is intended for operation on a x86–64-bit machine, running a 64-bit version of the operating system. It is recommended to use a machine with at least 1 TB of disk storage for one whole-genome sequencing data analysis and a minimum of 64GB of RAM. The novoBreak software supports multiple CPU cores. It is recommended to apply as many cores as possible on the machine.
2. Software setup. Download novoBreak binary distribution from github:

   $ git clone https://github.com/czc/nb_distribution.git

   If git tool is not installed, follow instructions on Git website: https://git-scm.com/.

   Add the downloaded package to your PATH environment variable:

   $ export PATH = $PWD/nb_distribution/:$PATH
3. Software dependency. The binary distribution depends on novoBreak core program for calculating novo k-mers, bwa-mem [20] (v0.7.10-r806-dirty and above) for contigs alignment, SSAKE [21] for local de novo assembly, and SAMtools [22] (v1.3 and above) for extracting reads. They have been

included in the binary distribution. Please test the dependencies first and install as necessary as follows:

(a) Test if "novoBreak" was installed properly by typing:

$ novoBreak

You should get command options shown as Fig. 2.

Otherwise, you should install novoBreak core program:

$ git clone https://github.com/czc/novobreak_src.git

$ cd novobreak_src && make; cd -

$ cp novobreak_src/novoBreak nb_distribution/

(b) Test if "bwa" was installed by typing:

$ bwa

You should see command options as shown in Fig. 3 or a higher version.

Otherwise, you should install bwa:

$ git clone https://github.com/lh3/bwa.git

$ cd bwa; make; cd -

$ cp bwa/bwa nb_distribution/

(c) Test if "SSAKE" was installed by typing:

$ SSAKE

You should notice command options shown as Fig. 4 or a higher version.

Otherwise, you should install SSAKE:

$ wget --no-check http://www.bcgsc.ca/platform/bioinfo/software/ssake/releases/3.8.5/ssake_v3-8-5.tar.gz

$ tar zxvf ssake_v3-8-5.tar.gz

$ cp ssake_v3.8.5/SSAKE nb_distribution/

(d) Test if "SAMtools" was installed properly:

$ samtools

You should see command options as shown in Fig. 5 or a higher version.

Otherwise, you should install SAMtools:

$ wget --no-check https://sourceforge.net/projects/samtools/files/samtools/1.3/samtools-1.3.tar.bz2/download -O samtools-1.3.tar.bz2

$ tar jxvf samtools-1.3.tar.bz2

$ cd samtools-1.3/ && make && cd -

$ cp samtools-1.3/samtools nb_distribution/

```
novoBreak - a tool for discovering somatic sv breakpoints
Auther: Zechen Chong <zchong@mdanderson.org>
Version: 1.1 (r20151007)
Usage:
  novoBreak -i <tumorbam> -c <normalbam> -r <reference> -o <output.kmer> [options]
Options:
  -h              This help
  -i <string>     Tumor bam file
  -c <string>     Normal bam file
  -r <string>     Reference file in fasta format
  -k <int>        Kmer size, <=31 [31]
  -o <string>     Output kmer
  -g <int>        Output germline events [0]
  -m <int>        Minimum kmer count regarded as novo kmers [3]
```

Fig. 2 novoBreak software command options

```
Program: bwa (alignment via Burrows-Wheeler transformation)
Version: 0.7.10-r806-dirty
Contact: Heng Li <lh3@sanger.ac.uk>

Usage:   bwa <command> [options]

Command: index         index sequences in the FASTA format
         mem           BWA-MEM algorithm
         fastmap       identify super-maximal exact matches
         pemerge       merge overlapping paired ends (EXPERIMENTAL)
         aln           gapped/ungapped alignment
         samse         generate alignment (single ended)
         sampe         generate alignment (paired ended)
         bwasw         BWA-SW for long queries

         fa2pac        convert FASTA to PAC format
         pac2bwt       generate BWT from PAC
         pac2bwtgen    alternative algorithm for generating BWT
         bwtupdate     update .bwt to the new format
         bwt2sa        generate SA from BWT and Occ

Note: To use BWA, you need to first index the genome with `bwa index'.
      There are three alignment algorithms in BWA: `mem', `bwasw', and
      `aln/samse/sampe'. If you are not sure which to use, try `bwa mem'
      first. Please `man ./bwa.1' for the manual.
```

Fig. 3 bwa software command options

3 Methods

We describe the detailed instructions of running novoBreak pipeline in this section. The interfaces of novoBreak pipeline require bam files. If you would like to analyze from raw FASTQ files, you need align the raw reads to the reference genome first and generate input bams for novoBreak workflow.

```
Usage: ./SSAKE [v3.8.1]
-f  File containing all the [paired (-p 1)] reads (required)
        With -p 1:
       ! Target insert size must be indicated at the end of the header line (e.g. :200 for a 200bp insert)
       ! Paired reads must be separated by ":"
        >template_name:200
        ACGACACTATGCATAAGCAGACGAGCAGCGACGCAGCACG:GCGCACGACGCAGCACAGCAGCAGACGAC
-w  Minimum depth of coverage allowed for contigs (e.g. -w 1 = process all reads [v3.7 behavior], required)
    *The assembly will stop when 50+ contigs with coverage < -w have been seen.*
-s  Fasta file containing sequences to use as seeds exclusively (specify only if different from read set, optional)
      -i Apply read space restriction to seeds (TASR behavior) while -s option in use (-i 1 = yes, default = no , optional)
-m  Minimum number of overlapping bases with the seed/contig during overhang consensus build up (default -m 20)
-o  Minimum number of reads needed to call a base during an extension (default -o 2)
-r  Minimum base ratio used to accept a overhang consensus base (default -r 0.7)
-t  Trim up to -t base(s) on the contig end when all possibilities have been exhausted for an extension (default -t 0, optional)
-c  Track base coverage and read position for each contig (default -c 0, optional)
-h  Ignore read name/header *will use less RAM if set to -h 1* (-h 1 = yes, default = no, optional)
-b  Base name for your output files (optional)
-z  Minimum contig size to track base coverage and read position (default -z 100, optional)
-p  Paired-end reads used? (-p 1 = yes, default = no, optional)
-v  Runs in verbose mode (-v 1 = yes, default = no, optional)
============ Options below only considered with -p 1 ============
-e  Error (%) allowed on mean distance   e.g. -e 0.75  == distance +/- 75% (default -e 0.75, optional)
-k  Minimum number of links (read pairs) to compute scaffold (default -k 4, optional)
-a  Maximum link ratio between two best contig pairs *higher values lead to least accurate scaffolding* (default -a 0.5, optional)
-x  Minimum overlap required between contigs to merge adjacent contigs in a scaffold (default -x 20, optional)
-n  N-pad gaps (-n 1 = yes, default = no 0, optional)
-g  Fasta file containing unpaired sequence reads (optional)
```

Fig. 4 SSAKE software command options

For example, given the raw tumor read files "tumor.read1.fq" and "tumor.read2.fq" and the reference "genome.fa" (*see* **Note 2**), you can align the reads to reference to generate the tumor reads alignment file:

$ bwa mem -T0 -t8 -p genome.fa tumor.read1.fq tumor.read2.fq | samtools view -Sb - | samtools sort -@8 -o tumor.bam - && samtools index tumor.bam

Here, we set 8 CPU threads to speed up the alignment (*see* **Note 3**). Similarly, given raw normal read files "normal.read1.fq" and "normal.read2.fq", you can generate the normal reads alignment file:

$ bwa mem -T0 -t8 -p genome.fa normal.read1.fq normal.read2.fq | samtools view -Sb - | samtools sort -@8 -o normal.bam - && samtools index normal.bam

3.1 Somatic SV Detection

The pipeline of novoBreak is written in Bash shell script. The command line option of novoBreak pipeline is shown in Fig. 6.

"<novoBreak_exe_dir>" indicates the path (*see* **Note 4**) to the binary distribution of the novoBreak pipeline.

"<ref>" indicates the path of the indexed reference base name.

"<tumor_bam>" indicates the path of the tumor bam file.

"<normal_bam>" indicates the path of the normal bam file.

"<n_CPUs:INT>" asks for setting the number of CPU cores to be used in the job.

"[outputdir:-PWD]", optional, indicates the path of the output directory. By default, the pipeline will write the output files in the current working directory.

Suppose that the two bam files are named as "tumor.bam" and "normal.bam", respectively. The reference is genome.fa and has been indexed using "bwa index" command. Assume that the

```
Program: samtools (Tools for alignments in the SAM format)
Version: 1.3 (using htslib 1.3)

Usage:   samtools <command> [options]

Commands:
  -- Indexing
     dict           create a sequence dictionary file
     faidx          index/extract FASTA
     index          index alignment

  -- Editing
     calmd          recalculate MD/NM tags and '=' bases
     fixmate        fix mate information
     reheader       replace BAM header
     rmdup          remove PCR duplicates
     targetcut      cut fosmid regions (for fosmid pool only)
     addreplacerg   adds or replaces RG tags

  -- File operations
     collate        shuffle and group alignments by name
     cat            concatenate BAMs
     merge          merge sorted alignments
     mpileup        multi-way pileup
     sort           sort alignment file
     split          splits a file by read group
     quickcheck     quickly check if SAM/BAM/CRAM file appears intact
     fastq          converts a BAM to a FASTQ
     fasta          converts a BAM to a FASTA

  -- Statistics
     bedcov         read depth per BED region
     depth          compute the depth
     flagstat       simple stats
     idxstats       BAM index stats
     phase          phase heterozygotes
     stats          generate stats (former bamcheck)

  -- Viewing
     flags          explain BAM flags
     tview          text alignment viewer
     view           SAM<->BAM<->CRAM conversion
     depad          convert padded BAM to unpadded BAM
```

Fig. 5 samtools software command options

```
./run_novoBreak.sh <novoBreak_exe_dir> <ref> <tumor_bam> <normal_bam> <n_CPUs:INT> [outputdir:-PWD]
```

Fig. 6 Command line option of novoBreak pipeline

available CPUs are 16 (*see* **Note 5**). Then you can use the following command:

$ bash run_novoBreak.sh /path/to/nb_distribution /path/to/genome.fa /path/to/tumor.bam /path/to/normal.bam 16 novoBreak_out

All the intermediate files and final output will be written to the directory “novoBreak_out”. The output files look like Fig. 7. The

```
total 5.1G
 34M -rw-rw-r-- 1 zchong  34M Mar 31 21:02 novoBreak.pass.flt.vcf
   0 drwxrwxr-x 3 zchong  512 Mar 31 21:01 ssake/
   0 drwxrwxr-x 3 zchong  512 Mar 31 17:41 group_reads/
128K -rw-rw-r-- 1 zchong  753 Mar 31 16:30 germlinereads.bam
3.7G -rw-rw-r-- 1 zchong 3.7G Mar 31 16:30 somaticreads.bam
1.5G -rw-rw-r-- 1 zchong 1.5G Mar 31 16:30 kmer.stat
```

Fig. 7 NovoBreak pipeline outputs

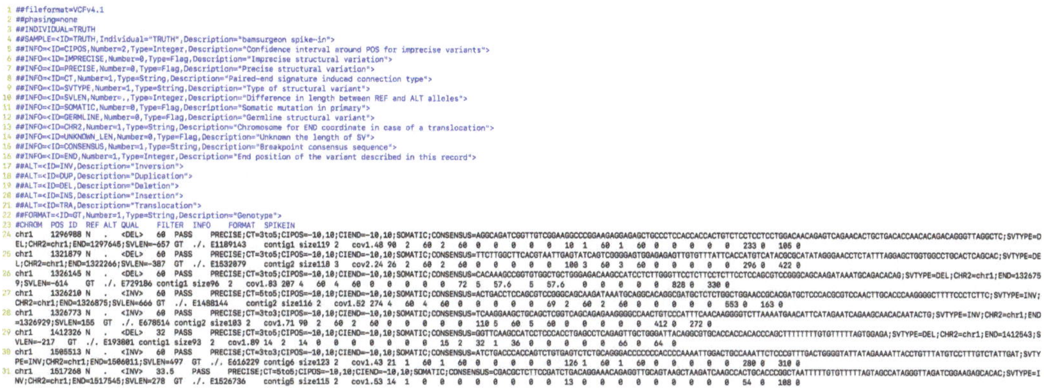

```
##fileformat=VCFv4.1
##phasing=none
##INDIVIDUAL=TRUTH
##SAMPLE=<ID=TRUTH,Individual="TRUTH",Description="bamsurgeon spike-in">
##INFO=<ID=CIPOS,Number=2,Type=Integer,Description="Confidence interval around POS for imprecise variants">
##INFO=<ID=IMPRECISE,Number=0,Type=Flag,Description="Imprecise structural variation">
##INFO=<ID=PRECISE,Number=0,Type=Flag,Description="Precise structural variation">
##INFO=<ID=CT,Number=1,Type=String,Description="Paired-end signature induced connection type">
##INFO=<ID=SVTYPE,Number=1,Type=String,Description="Type of structural variant">
##INFO=<ID=SVLEN,Number=.,Type=Integer,Description="Difference in length between REF and ALT alleles">
##INFO=<ID=SOMATIC,Number=0,Type=Flag,Description="Somatic mutation in primary">
##INFO=<ID=GERMLINE,Number=0,Type=Flag,Description="Germline structural variant">
##INFO=<ID=CHR2,Number=1,Type=String,Description="Chromosome for END coordinate in case of a translocation">
##INFO=<ID=UNKNOWN_LEN,Number=0,Type=Flag,Description="Unknown the length of SV">
##INFO=<ID=CONSENSUS,Number=1,Type=String,Description="Breakpoint consensus sequence">
##INFO=<ID=END,Number=1,Type=Integer,Description="End position of the variant described in this record">
##ALT=<ID=INV,Description="Inversion">
##ALT=<ID=DUP,Description="Duplication">
##ALT=<ID=DEL,Description="Deletion">
##ALT=<ID=INS,Description="Insertion">
##ALT=<ID=TRA,Description="Translocation">
##FORMAT=<ID=GT,Number=1,Type=String,Description="Genotype">
#CHROM  POS ID  REF ALT QUAL    FILTER  INFO    FORMAT  SPIKEIN
chr1    1296988 N   .   <DEL>   60  PASS    PRECISE;CT=3to5;CIPOS=-10,10;CIEND=-10,10;SOMATIC;CONSENSUS=AGGCAGATCGGTTGTCGGAAGGCCCGGAAGAGGAGAGCTGCCCTCCACCACCACTGTCTCCTCCTCCTGGACAACAGAGTCAGAACACTGCTGACACCAACACAGACAGGGTTAGGCTC;SVTYPE=D
EL;CHR2=chr1;END=1297645;SVLEN=-657 GT  ./. E1189143    contig1 size119 2   cov1.48 90  2   60  2   60  0   0   0   0   0   10  1   60  1   60  0   0   0   0   0   233 0   105 0
chr1    1321879 N   .   <DEL>   60  PASS    PRECISE;CT=3to5;CIPOS=-10,10;CIEND=-10,10;SOMATIC;CONSENSUS=TTCTTGGCTTCACGTAATTGAGTATCAGTCGGGGAGTGGAGAGAGTTGTGTTTATTCACCATGTCATACGCGCATATAGGGAACCTCTATTTAGGAGCTGGTGGCCTGCACTCAGCAC;SVTYPE=DE
L;CHR2=chr1;END=1322266;SVLEN=-387  GT  ./. E1532079    contig2 size118 3   cov2.24 26  2   60  2   60  0   0   0   0   0   100 3   60  3   60  0   0   0   0   0   296 0   422 0
chr1    1326145 N   .   <DEL>   60  PASS    PRECISE;CT=3to5;CIPOS=-10,10;CIEND=-10,10;SOMATIC;CONSENSUS=CACAAAGCCGGTGTGGCTGCTGGGAGACAAGCCATCCTCTTGGGTTCCTCTTCCTCTTCCTCCAGCGTCCGGGCAGCAAGATAAATGCAGACACAG;SVTYPE=DEL;CHR2=chr1;END=132675
9;SVLEN=-614    GT  ./. E729186 contig1 size96  2   cov1.83 207 4   60  4   60  0   0   0   0   0   72  5   57.6    5   57.6    0   0   0   0   0   828 0   330 0
chr1    1326210 N   .   <INV>   60  PASS    PRECISE;CT=5to5;CIPOS=-10,10;CIEND=-10,10;SOMATIC;CONSENSUS=ACTGACCTCCAGCGTCCGGGCAGCAAGATAAATGCAGGCACAGGCGATGCTCTCTGGCTGGAACCGCACGATGCTCCCACGCGTCCAACTTGCACCCAAGGGGCTTTTCCCTCTTC;SVTYPE=INV;
CHR2=chr1;END=1326875;SVLEN=666 GT  ./. E1488144    contig2 size116 2   cov1.52 274 4   60  4   60  0   0   0   0   0   69  2   60  2   60  0   0   0   0   0   553 0   163 0
chr1    1326773 N   .   <INV>   60  PASS    PRECISE;CT=3to3;CIPOS=-10,10;CIEND=-10,10;SOMATIC;CONSENSUS=TCAAGGAAGCTGCAGCTCGGTCAGCAGAGAAGGGGCCAACTGTCCCATTTCAACAAGGGGTCTTAAAATGAACATTCATAGAATCAGAAGCAACACAATACTG;SVTYPE=INV;CHR2=chr1;END
=1326929;SVLEN=155  GT  ./. E678514 contig2 size103 2   cov1.71 90  2   60  2   60  0   0   0   0   0   110 5   60  5   60  0   0   0   0   0   412 0   272 0
chr1    1412326 N   .   <DEL>   32  PASS    PRECISE;CT=3to5;CIPOS=-10,10;CIEND=-10,10;SOMATIC;CONSENSUS=GGTTCAAGCCATCCTCCCACCTGAGCCTCAGAGTTGCTGGGATTACAGGCGTGCACCACCACACCCAGCTTTTTTTTGTGTTTTTAGTGGAGA;SVTYPE=DEL;CHR2=chr1;END=1412543;S
VLEN=-217   GT  ./. E193801 contig1 size93  2   cov1.89 14  2   14  0   0   0   0   0   0   0   15  2   32  1   36  0   0   0   0   0   66  0   64  0
chr1    1505513 N   .   <INV>   60  PASS    PRECISE;CT=3to3;CIPOS=-10,10;CIEND=-10,10;SOMATIC;CONSENSUS=ATCTGACCCACCAGTCTGTGAGTCTCTGCAGGGACCCCCCACCCCAAAATTGGACTGCCAAATTCTCCCGTTTGACTGGGGTATTATAGAAAATTACCTGTTTATGTCCTTTGTCTATTGAT;SVTY
PE=INV;CHR2=chr1;END=1506011;SVLEN=497  GT  ./. E616229 contig6 size123 2   cov1.43 21  1   60  1   60  0   0   0   0   0   126 1   60  1   60  0   0   0   0   0   280 0   310 0
chr1    1517268 N   .   <INV>   33.5    PASS    PRECISE;CT=5to5;CIPOS=-10,10;CIEND=-10,10;SOMATIC;CONSENSUS=CGACGCTCTTCCGATCTGACAGGAAACAGAGGTTGCAGTAAGCTAAGATCAAGCCACTGCACCCGGCTAATTTTTGTGTTTTTAGTAGCCATAGGGTTAGATCGGAAGAGCACAC;SVTYPE=I
NV;CHR2=chr1;END=1517545;SVLEN=278  GT  ./. E1526736    contig5 size115 2   cov1.53 14  1   0   0   0   0   0   0   0   0   13  0   0   0   0   0   0   0   0   0   0   54  0   108 0
```

Fig. 8 The format of novoBreak final output file

output file "novoBreak.pass.flt.vcf" contains filtered results (*see* Fig. 8 for an example). It follows the standard VCF file format and adds a few more fields for filtering (*see* **Note 6**) purpose.

The pipeline can always guarantee a good sensitivity. The file "ssake/ssake.pass.vcf" contains a highly sensitive call set. After inspecting the bam files, a filter will be applied to the inspected VCF files (*.sp.vcf). The default filter may not be optimal due to the detailed sequencing experiment. You can develop your own filters to apply to your data (*see* **Note 7**).

3.2 Germline SV Detection

To detect germline SVs, you should provide a mocked "normal" bam file to meet the interfaces of novoBreak pipeline.

You can use the simulator, wgsim, in SAMtools [22] package:

$ wget --no-check https://sourceforge.net/projects/samtools/files/samtools/1.3/samtools-1.3.tar.bz2/download -O samtools-1.3.tar.bz2

$ tar jxvf samtools-1.3.tar.bz2

$ cd samtools-1.3/ && make && cd -

$ cp samtools-1.3/misc/wgsim nb_distribution/

To simulate the normal reads from a reference sequence "genome.fa":

$ wgsim -e 0.001 -1 100 -2 100 -r 0 genome.fa normal.read1.fq normal.read2.fq

Then, you can follow the instructions of previous section to generate the normal bam file and execute the pipeline, respectively.

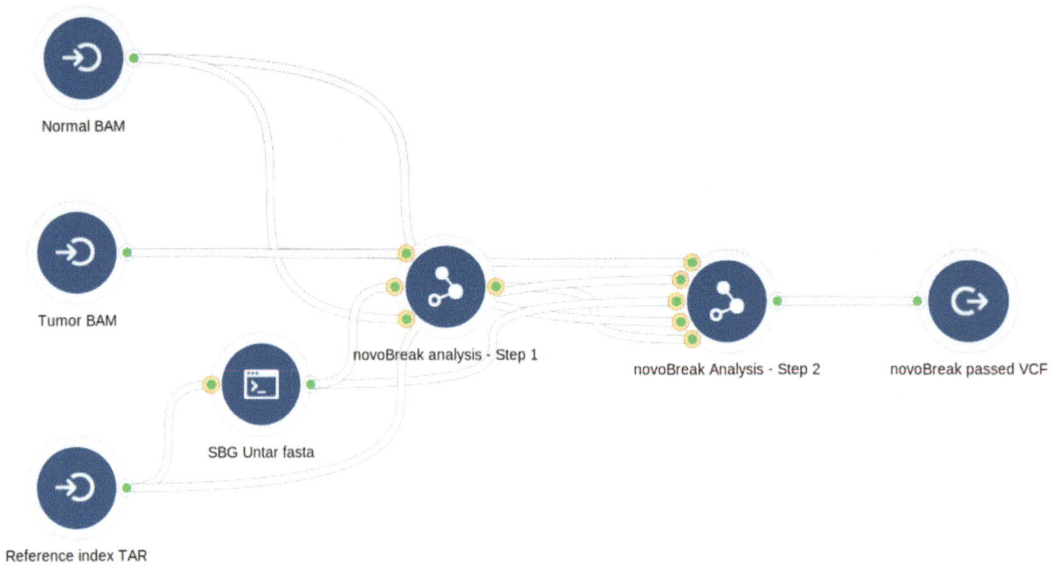

Fig. 9 The workflow of novoBreak pipeline on Cancer Genomics Cloud platform

You may need to apply a different filter for germline SV detection. The "SOMATIC" label in the "novoBreak.pass.flt.vcf" should change to "GERMLINE":

$ sed 's/SOMATIC/GERMLINE/' novoBreak.pass.flt.vcf > novoBreak.pass.germline.flt.vcf

3.3 Cancer Genomics Cloud of Seven Bridges

The Seven Bridges Bioinformatics team and we have optimized novoBreak pipeline on the Cancer Genomics Cloud (CGC) platform (https://cgc.sbgenomics.com/public/apps#ZCHONG/novobreak-commit/novobreak-analysis/). The cost and running time have been reduced dramatically, which can meet the need for large scale analysis of tumor–normal paired samples on the Cloud environment. The workflow of novoBreak works as shown in Fig. 9. Similar to local machine pipeline, the inputs require an indexed tumor bam file, an index normal bam file and an indexed reference. Different to local machine pipeline, the reference and its indexes need to be put in a TAR bundle file. To run tasks on CGC platform, you need to follow instructions on CGC (http://docs.cancergenomicscloud.org/docs) to sign up a CGC account, to gain access to protected data (TCGA for example), etc.

4 Notes

1. For tumor–normal pair studies (*see* Subheading 3.1), the two bam files are tumor bam and normal bam, respectively. For a single sample study (*see* Subheading 3.2), the "tumor" bam file

should be the sample and the "normal" bam file should be the simulated bam. All the bam files should be sorted by coordinates and indexed using SAMtools or Picard.

2. The reference should be indexed using "bwa index" and "samtools faidx" commands. After indexing, you should see files with names "genome.fa, genome.fa.amb, genome.fa.ann, genome.fa.bwt, genome.fa.fai, genome.fa.pac, genome.fa.sa". The reference should be the same version as the bam file header shows (using command lines "samtools view -H $file.bam" and "cat $reference.fai" to check).
3. You should set the number of threads based on the capability of the machine.
4. You only need to provide a relative path indicating the directory. The novoBreak pipeline will calculate the absolute path.
5. Note that not all the processes (shell commands) in the novoBreak pipeline will be executed with the allocated CPUs. Only the local de novo assembly process ("run_ssake"), alignment ("bwa mem"), and retrospect breakpoints ("infer_bp") will be run in parallel. You may need to separate the pipeline into five parts in some cases. For example, for the git committed version with HSA-1 checksum '22b4a3155be39871f472627fec02eb52ddc3866a', part 1 is from line 1 to line 39 which uses 1 CPU, part 2 from line 40 to line 53 which can be paralleled, part 3 from line 54 to line 62 which uses 1 CPU, part 4 from line 63 to line 69 which can be paralleled again, and part 5 from line 70 to the end which only uses 1 CPU. For other versions, please adjust the parts accordingly if the pipeline is different.
6. Besides the standard 10 fields, the VCF file contains 29 additional fields can be used to build your own filters. These fields (column 11 to column 39) are cluster_id, contig_id, contig_size, reads_used_for_assembly, average_coverage, tumor_bkpt1_depth, tumor_bkpt1_sp_reads, tumor_bkpt1_qual, tumor_bkpt1_high_qual_sp_reads, tumor_bkpt1_high_qual_qual, normal_bkpt1_depth, normal_bkpt1_sp_reads, normal_bkpt1_qual, normal_bkpt1_high_qual_sp_reads, normal_bkpt1_high_qual_qual, tumor_bkpt2_depth, tumor_bkpt2_sp_reads, tumor_bkpt2_qual, tumor_bkpt2_high_qual_sp_reads, tumor_bkpt2_high_qual_qual, normal_bkpt2_depth, normal_bkpt2_sp_reads, normal_bkpt2_qual, normal_bkpt2_high_qual_sp_reads, normal_bkpt2_high_qual_qual, tumor_bkpt1_discordant_reads, normal_bkpt1_discordant_reads, tumor_bkpt2_discordant_reads, normal_bkpt2_discordant_reads. Here, "bkpt" represents "breakpoint"; "sp" represents "split"; "qual" stands for "quality".

7. To increase sensitivity, novoBreak tries to infer as many SVs as possible based on the local assembly results. But many of the inferred SVs may be false positives due to misassembly or lack of enough evidence. So we provided a default filter to get a relatively stringent filtered call set based on real data experience. We empirically defined the minimum SV size as 100 bp and no upper limit. Users can change the filter and cutoffs based on the utility and the knowledge as needed. An empirical filter can be made based on the column 6 of novoBreak's output. A higher value of column 6 indicates a more reliable event. The field descriptions in **Note 6** should also be considered to provide a sensitive filter.

References

1. Kloosterman WP, Francioli LC, Hormozdiari F et al (2015) Characteristics of de novo structural changes in the human genome. Genome Res 25:792–801
2. Berger MF, Lawrence MS, Demichelis F et al (2011) The genomic complexity of primary human prostate cancer. Nature 470:214–220
3. Hillmer AM, Yao F, Inaki K et al (2011) Comprehensive long-span paired-end-tag mapping reveals characteristic patterns of structural variations in epithelial cancer genomes. Genome Res 21:665–675
4. Campbell PJ, Yachida S, Mudie LJ et al (2010) The patterns and dynamics of genomic instability in metastatic pancreatic cancer. Nature 467:1109–1113
5. Mertens F, Johansson B, Fioretos T, Mitelman F (2015) The emerging complexity of gene fusions in cancer. Nat Rev Cancer 15:371–381
6. Chen K, Wallis JW, McLellan MD et al (2009) BreakDancer: an algorithm for high-resolution mapping of genomic structural variation. Nat Methods 6:677–681
7. Hormozdiari F, Alkan C, Eichler EE, Sahinalp SC (2009) Combinatorial algorithms for structural variation detection in high-throughput sequenced genomes. Genome Res 19:1270–1278
8. Abyzov A, Urban AE, Snyder M, Gerstein M (2011) CNVnator: an approach to discover, genotype, and characterize typical and atypical CNVs from family and population genome sequencing. Genome Res 21:974–984
9. Ye K, Schulz MH, Long Q et al (2009) Pindel: a pattern growth approach to detect break points of large deletions and medium sized insertions from paired-end short reads. Bioinformatics 25:2865–2871
10. Hajirasouliha I, Hormozdiari F, Alkan C et al (2010) Detection and characterization of novel sequence insertions using paired-end next-generation sequencing. Bioinformatics 26:1277–1283
11. Rausch T, Zichner T, Schlattl A et al (2012) DELLY: structural variant discovery by integrated paired-end and split-read analysis. Bioinformatics 28:i333–i339
12. Chen K, Chen L, Fan X et al (2014) TIGRA: a targeted iterative graph routing assembler for breakpoint assembly. Genome Res 24:310–317
13. Alkan C, Coe BP, Eichler EE (2011) Genome structural variation discovery and genotyping. Nat Rev Genet 12:363–376
14. Medvedev P, Stanciu M, Brudno M (2009) Computational methods for discovering structural variation with next-generation sequencing. Nat Methods 6:S13–S20
15. Stephens PJ, Greenman CD, Fu B et al (2011) Massive genomic rearrangement acquired in a single catastrophic event during cancer development. Cell 144:27–40
16. Baca SC, Prandi D, Lawrence MS et al (2013) Punctuated evolution of prostate cancer genomes. Cell 153:666–677
17. Li Y, Zheng H, Luo R et al (2011) Structural variation in two human genomes mapped at single-nucleotide resolution by whole genome de novo assembly. Nat Biotechnol 29:723–730
18. Earl D, Bradnam K, John JS, Darling A, Lin D, Fass J, Yu HOK, Buffalo V, Zerbino DR, Diekhans M et al (2011) Assemblathon 1: A competitive assessment of de novo short read assembly methods. Genome Res 21:2224–2241

19. Chong Z, Ruan J, Gao M et al (2017) novoBreak: local assembly for breakpoint detection in cancer genomes. Nat Methods 14:65–67
20. Li H (2013) Aligning sequence reads, clone sequences and assembly contigs with BWA-MEM. arXiv:1303.3997 [q-bio.GN]
21. Warren RL, Sutton GG, Jones SJM, Holt RA (2007) Assembling millions of short DNA sequences using SSAKE. Bioinformatics 23:500–501
22. Li H, Handsaker B, Wysoker A et al (2009) The sequence alignment/map format and SAMtools. Bioinformatics 25:2078–2079

Chapter 11

Use of RAPTR-SV to Identify SVs from Read Pairing and Split Read Signatures

Derek M. Bickhart

Abstract

High-throughput short read sequencing technologies are still the leading cost-effective means of assessing variation in individual samples. Unfortunately, while such technologies are eminently capable of detecting single nucleotide polymorphisms (SNP) and small insertions and deletions, the detection of large copy number variants (CNV) with these technologies is prone to numerous false positives. CNV detection tools that incorporate multiple variant signals and exclude regions of systemic bias in the genome tend to reduce the probability of false positive calls and therefore represent the best means of ascertaining true CNV regions. To this end, we provide instructions and details on the use of the RAPTR-SV CNV detection pipeline, which is a tool that incorporates read-pair and split-read signals to identify high confidence CNV regions in a sequenced sample. By combining two different structural variant (SV) signals in variant calling, RAPTR-SV enables the easy filtration of artifact CNV calls from large datasets.

Key words Read pair, Split-read, Combined detection, RAPTR-SV, Whole genome sequencing

1 Introduction

Copy number variants (CNV) represent large (>50 bp) fluctuations of gene structure within individuals of a population. The detection of CNVs within sequenced individuals is quite difficult, owing to the larger sizes of many CNVs and the shortness of the sequencing reads used to detect them. Additionally, the use of single-copy (haploid) reference genome assemblies to detect complex, multi-copy events can preclude the interpretation of aligned sequencing data into copy number states. As high-throughput sequencing reads tend to be small and fragmentary, it was quickly realized that the distance and order of paired-end read fragments could be used as a proxy determinant for CNVs and structural variants (SV) [1]. Initially, programs designed for this purpose were quite adept at discovering deletions and insertions of sequence based on read insert length deviations in sets of "discordant" read pairs that had predicted insert lengths greater than or less than the average read

Derek M. Bickhart (ed.), *Copy Number Variants: Methods and Protocols*, Methods in Molecular Biology, vol. 1833,
https://doi.org/10.1007/978-1-4939-8666-8_11, © Springer Science+Business Media, LLC, part of Springer Nature 2018

pair in the sequencing library [2–4]. Further modification of paired end SV analysis led to the design of "split read" algorithms, which attempted to realign portions of the unmapped portion of a read pair that has one mapped read in the pair in order to determine fine-scale breakpoints of SV regions [5]. Both methods were suitable for SV discovery and had the benefit of being agnostic to sample read depth—a major limitation of alternative read-depth based CNV detection methods. However, spurious sequence alignment and chimeric reads from faulty library preparation remain a major source of bias in any detection method involving paired-end reads. A major breakthrough resulted from the use of both methods of SV discovery in a combined fashion to reduce the number of false positives due to these methods of bias [6, 7]. Still, these combined methods of detection are most useful on samples that have highly polished reference genomes relatively devoid of gap sequence and misassemblies.

In order to include all aspects of read pairing in the determination of SV and CNV calls in the genomes of nonmodel organisms, we developed RAPTR-SV as a pipeline method [8]. RAPTR-SV is written in the Java programming language and is designed to use multiple processor cores to speed analysis. RAPTR-SV works by interrogating all discordant paired-end reads from a BWA-aligned [9] sequence map file [10], and collects one-end anchor reads for subsequent split read analysis. After all one-end anchor reads have been collected, the unmapped portion of the read pair is split and aligned to the reference genome using MrsFAST [11] to generate split read information. RAPTR-SV then collects all of the discordant and split read data into nonoverlapping sets in order to determine a consensus SV or CNV call within a specific region. Given the differences in SV resolution provided by each class of variant read pair, RAPTR-SV uses discordant read pairs to identify general SV regions and uses split read pairs to determine SV breakpoints. Additionally, RAPTR-SV incorporates reference genome gap information and other quality metrics designed to aid SV calling in nonfinished reference genomes. The results are an accurate assessment of structural incongruity in an analyzed sample.

2 Materials

In order to install and run RAPTR-SV, you must install two prerequisite software packages. Currently, RAPTR-SV works only in a Unix-based environment given its dependency on MrsFAST [11].

2.1 Installing Java Version 8

You must first install Java version 1.8+ by downloading and running an installation package downloaded from the following URL: www.java.com/en/

In order to invoke Java, you simply type the java command on the command line, followed by several instructions:

$ java -Xmx(Z)g –jar (jar file)

Where "Z" is the number of gigabytes of RAM needed by the program (*see* **Note 1**) and the "jar file" is a Java Archive (JAR) file containing the source code of your program. If you do not specify the gigabytes of RAM for your java program, the system will dedicate a set amount (typically 25% of the total system memory) by default. If the program exceeds that memory limit, the java virtual machine (JVM) will terminate itself with an error.

2.2 Installing MrsFAST-Ultra

RAPTR-SV uses MrsFAST [11, 12] as an alignment engine for split read alignments, so its installation and inclusion on the user's PATH variable is a requirement for use of RAPTR-SV (*see* **Note 2**). In order to download MrsFAST-Ultra, you can either download a precompiled binary from the developer's github repository (https://github.com/sfu-compbio/mrsfast/releases/tag/v3.4.0) or you can download the repository itself and compile the binaries:

$ git clone https://github.com/sfu-compbio/mrsfast.git

$ cd mrsfast

$ make; make install

The "mrsfast" binary must be on the user PATH in order to run RAPTR-SV preprocessing mode. In order to test if MrsFast is currently installed on the PATH, type the following command:

$ which mrsfast

If a directory for the program is given, then MrsFast is correctly installed. If there is an error or the program is not found on your PATH, then you may add it to the PATH variable using the following command:

$ export PATH=$PATH:/your/directory/to/MrsFAST/

Typing a "which mrsfast" should then give you the correct location of the MrsFast executable.

3 Methods

RAPTR-SV has two modes of operation: (1) to preprocess sequence read alignment files (BAM format) [10] to extract discordant read pairs and (2) to generate a consensus structural variant call from the combined information present in the identified discordant reads. The modes must be run in order from the "preprocess" mode to the "cluster" mode, respectively, with input file formats specific to each mode. However, there are universal command line arguments that the user can present to RAPTR-SV to control the flow and pace of analysis.

Table 1
RAPTR-SV command line arguments common to both modes

Argument	Value	Description	Default
-p	String	Alternative temporary file directory	/tmp or the $TEMP folder
-t	Integer	Number of threads to use	A single thread
-d	(none)	Debug mode	Disabled by default

3.1 Common Command Line Arguments and Working Space

Both of the RAPTR-SV modes accept a set of common command line arguments (*see* Table 1). Depending on your system architecture and working space setup, setting an alternative temporary file directory is likely to be the most important common command issued to each RAPTR-SV process. Given the large amount of information that RAPTR-SV stores on the hard disk, you may need in excess of 500 megabytes of disk storage to process BAM files containing approximately 30X coverage sequence reads (*see* **Note 3**). It is highly recommended that you create a separate temporary directory in your system workspace and that you set the –p command line option to point to this directory every time you run the program.

Regarding the other two command options, their use is encouraged but they may not always be necessary in all use cases. RAPTR-SV has been designed to include several java v1.8 stream paradigms for faster processing, so many portions of the "preprocess" and "cluster" algorithm take advantage of multicore processors. Since not all portions of the program could be threaded, performance is not increased linearly based on the number of threads passed to the program [8]. Still, RAPTR-SV generates data faster as more threads are added via the –t argument. Finally, the debug flag, −d, is designed to provide more verbose log output to troubleshoot RAPTR-SV errors. RAPTR-SV generates independent log files for each runtime and mode, enabling the user to assess program performance and to quickly spot errors. If an error message is particularly cryptic, use of the debug flag will provide more information for use in troubleshooting or to inform the software developers of existing bugs.

3.2 Preprocessing Mode

The first step in the RAPTR-SV pipeline is to extract read insert statistics, discordant read pairs and split-reads using the "preprocess" mode utility. RAPTR-SV preprocess mode has several required and optional input parameters (*see* Table 2) with optional parameters designed around partitioning and filtering of data. In most use cases, the default RAPTR-SV options are sufficient for generating high quality SV calls. Users may tweak these parameters

Table 2
RAPTR-SV preprocess mode arguments

Argument	Value	Description	Default
-i	String	The input BAM file	Required
-o	String	The base name for output files	Required
-r	String	A MrsFAST-indexed reference genome	Required
-g	(none)	Ignore BAM file read groups	Each read group is processed separately
-m	Integer	Maximum discordant read insert length	Pairs with an insert length over 1,000,000 bp are discarded
-s	Integer	Metadata sampling limit	Reads the first 10,000 read pairs to estimate insert length distributions

to suit other analyses or to overcome biases in their prepared data. For example, it may be useful to increase the metadata sampling limit ("-s") to over 1,000,000 to improve the accuracy of read pair insert length estimates. For cancer cell lines, or samples that are subject to numerous, large structural variants via chromothripsis [13], the read insert length threshold ("-m") may be set to a particularly large number (300,000,000) to disable this filter. For normal use cases (i.e., resequenced tissues from a multicellular eukaryote), the default parameters tend to give very accurate results and a minimum of false-positive SV calls.

The main carryover output from the "preprocess" mode of RAPTR-SV is a text file with the extension ".flat," henceforth referred to as a "flat file." This flat file contains the directory paths to the discordant read pair files and it also contains summaries of read length statistics for each read group present in the input BAM file (*see* **Note 4**). If you use the unix "cat" command on this file, you can view each read group's insert length statistical summaries which may assist you in identifying read groups that are full of highly variable insert length sequence reads. If you notice that a read group has undesirable read insert statistics (or you wish to proceed without that library for any reason), removing the offending read group prior to SV calling is as simple as deleting the line that contains its information in the flat file. The flat file is the main input into the second and last RAPTR-SV mode, where copy number variants and structural variants are called.

3.3 Cluster Mode

After running preprocess mode, the discordant read information is packaged into data file formats and structures useable by RAPTR-SV's cluster mode. In order to call SVs, one only needs to invoke RAPTR-SV cluster mode with the "-s" and "-o" arguments specifying

Table 3
RAPTR-SV cluster mode arguments

Argument	Value	Description	Default
-s	String	The flat file output from RAPTR-SV "preprocess" mode	Required
-o	String	The base name for cluster mode output files	Required
-c	String	Cluster only this chromosome	Cluster each chromosome present in the RAPTR-SV preprocess flat file output
-g	String	Assembly gap file with intervals to discard spurious read alignments	Do not discard reads that intersect with assembly gaps
-m	Decimal	Set the mapping quality filter for discordant reads	Pairs with a quality less than 0.0001 are discarded
-f	Decimal	Set the mapping quality weight threshold to call true SVs	Any SV with a cumulative mapping quality weight greater than 1.00 is called
-i	Integer	Set the raw read count for SV calls	Any SV with fewer than two reads is discarded
-b	Integer	The number of SV calls to hold in memory at any time	Up to 10 SV sets are held in memory at any time.

the input flat file and the output call file base name, respectively (*see* Table 3). At the conclusion of the cluster mode program, the user will be presented with three text files containing detected insertions (".raptr.insertions"), deletions (".raptr.deletions"), and tandem duplications (".raptr.tand"), respectively. The output file formats for each event are consistent with each other, and display the uncertainty of the SV event in both raw metrics and statistics (*see* Table 4). If the SV event is supported by several split reads, one can use the internal start and end coordinates (columns 3 and 4) to sort SVs by their location in the genome. Otherwise, the external and internal coordinates will be identical pairs (e.g., columns 2 and 3, and columns 4 and 5 will be identical in each pair, respectively) owing to the paucity of read pair information in the area of the SV.

It is possible to adjust the parameters of RAPTR-SV's cluster mode in order to reduce the number of false positive SV calls. It is highly recommended that users pass an assembly gap file to the program using the "-g" flag to remove SVs associated with faulty alignments to large repetitive structures that flank assembly gap regions. Many reference genome databases provide gap location files; however, one may generate a file from a new reference assembly using the BedTools suite [14] or through the use of a specialized program (https://github.com/njdbickhart/GetMaskBedFasta). The "-m" and "-f" flags remove discordant

Table 4
Cluster mode output file format

Column number	Brief description
1	Chromosome
2	5′ outer start
3	5′ internal start (SV breakpoint start)
4	3′ internal end (SV breakpoint end)
5	3′ outer end
6	SV prediction type
7	Number of supporting discordant read pairs
8	Number of supporting balanced split read pairs
9	Number of supporting unbalanced split read pairs
10	The total weight support of all reads supporting this event

read pairs that are likely to be due to repetitive alignments and SVs that are mostly comprised of read pairs due to repetitive alignments, respectively. For more information on the probability estimate used to determine this value, please see the work of Hormozdiari et al. [15]. Most users will not need to adjust this threshold to filter SV calls; however, for reference genome assemblies with high occurrence repetitive content (many elements with >60 copies), it may be worthwhile to set the "-m" and "-f" filters to lower values. Finally, it is highly recommended that the "-i" raw read count filter is adjusted to at least 1/3rd of the read coverage of your input bam (e.g., 30X read depth would have a recommended "-i" value of "10"). This enables the detection of heterozygous duplications and deletions while preventing single discordant or chimeric read pairs from constituting a valid SV call.

3.4 Optimizing Performance

RAPTR-SV tags and tracks a large proportion of variant reads from any input BAM file, so a large amount of system resources is needed to call SVs on each input sample. In order to optimize the use of server resources, we highly recommend the adjustment of several parameters to ensure that memory and disk usage is kept to a minimum. We always recommend the use of the alternative temporary directory ("-p") and the use of a large number of threads for each RAPTR-SV mode ("-t"). Temporary file directories can get fairly congested, so it is recommended that the directories are emptied after a program "crash." For the "preprocess" mode, the only additional tuning parameter is the metadata sampling limit ("-s") which is also used to determine how many read pairs are stored in memory at any specific time. Lower values reduce memory overhead but may increase processing time. For the "cluster" mode, it is highly recommended that the user include an assembly gap file ("-g") and adjust the raw read threshold ("-i") to an appropriate value. Both of these filters reduce the number of spurious SV calls that RAPTR-SV must consider in each calling iteration and therefore reduce the overall time needed to make SV calls. Finally, if memory overhead is still an issue while running "cluster" mode, we recommend adjusting the "-b" flag to a lower number, as this represents the number of read pairs per SV set that are held in memory at any time.

3.5 Filtering Output and Validating Calls

After generating SV calls using any software package, you are often left with the daunting task of validating your callset and identifying biologically relevant data. Given the number and types of SV calls from a typical RAPTR-SV survey, it is best to focus on the easily identifiable calls and to remove putative false positive events before they are included in your association analysis. The first priority for your analysis is to remove the top likely causes for false positives in paired-end SV analysis: (1) misalignment of reads to repetitive elements and (2) chimeric read alignments. RAPTR-SV enables the automatic filtration of many of these events by allowing the user to insert an assembly gap file ("-g" in cluster mode) and to set the weight of raw reads for SV assignment ("-i" in cluster mode). In the case of misalignment to repetitive reads, assembly gaps tend to exist within large repetitive elements in older reference genomes [16], so the filtration of any SV that spans an assembly gap will tend to remove a large portion of misaligned repetitive read pairs. Assuming proper library preparation protocols are followed, the proportion of chimeric read fragments in a sequencing library should be low and randomly distributed. By setting the threshold of the number of raw reads higher in the "-i" filter, you may automatically remove SV calls resulting from singleton chimeric read fragments. We have provided a sample workflow for running aligned reads in RAPTR-SV to take advantage of these automatic filters (*see* Table 5).

Even after automatic filtration, putative SV callsets may number in the tens of thousands. In order to filter RAPTR-SV datasets post hoc, we have included a filtration script as part of a separate Github repository (https://github.com/njdbickhart/perl_toolchain). The script calculates the average depth of coverage of the original BAM file and uses that coverage estimate to select variant calls that have a coverage that is consistent with heterozygous, homozygous or multiple-copy variant states. It also filters reads

Table 5
Example RAPTR-SV calling pipeline

Commands
Index reference
$ mrsfast –index /my/mrsfast/indexed/reference.fa
Identify gaps
$ java –jar GetMaskBedFasta.jar –f /my/mrsfast/indexed/reference.fa –o /my/mrsfast/indexed/reference.gaps.bed –s /my/mrsfast/indexed/reference.gaps.stats
Preprocess
$ java –Xmx50g –jar RAPTR-SV.jar preprocess –p /my/temp –t 5 –i /my/indexed/bam.bam –o /my/output/files –r /my/mrsfast/indexed/reference.fa -g
Cluster
$ java –Xmx50g –jar RAPTR-SV.jar cluster –p /my/temp –t 5 –s /my/output/files.flat –o /my/output/calls –g /my/mrsfast/indexed/reference.gaps.bed
Filtering
$ perl /perl_toolchain/sequence_data_scripts/filterRAPTRSVFiles.pl –b /my/indexed/bam.bam –i /my/output/calls.deletions –o /my/output/calls.deletions.filtered

based on the support of unbalanced and balanced split-reads (*see* **Note 5**) as well as discordant read pairs similar to standards advocated by Hormozdiari et al. [15]. Finally, it may be useful to associate SV calls with genomic annotations (i.e., gene regions) to identify priority SVs for further study. By using the BEDTools suite [14] or a separate annotation utility (https://github.com/njdbickhart/AnnotateUsingGenomicInfo), it is possible to intersect SV calls with biological information to characterize the putative functional impacts of your SV callset.

4 Notes

1. It is highly recommended that you allocate at least 50 gigabytes of RAM for an average X coverage (30X) aligned sample with a genome size greater than 2 gigabases. Memory requirements vary based on the number of discordant reads in your alignment file!
2. Older versions of MrsFAST may be used; however, the newest versions of the alignment program can use parallel processing and are recommended.
3. It is always good practice to maintain a large amount of working space; however, RAPTR-SV uses more working space than most programs. Having an excess of 1 terabyte of free space is

a good idea to ensure that the preprocess mode completes successfully.

4. The specific files produced by RAPTR-SV preprocess mode include: (1) a discordant read pair map file (*.divet), (2) a BAM file with split-read anchor reads (*.anchor.bam), (3) a BAM file with split-read mappings (*.bam) and (4) a tab-delimited text file that gives the full system paths to these files (*.flat). The separate mapping files are suitable for use in other software programs or as part of a separate analysis. Since all discordant reads and split reads are output to their respective files, the user is able to test for trans-chromosomal alignments indicative of mobile element insertions or Robertsonian translocations.

5. Unbalanced split-reads are split-read alignments where one half of the read successfully aligns and the other cannot (as it spans the SV site). Balanced split-reads are alignments where both split-read halves successfully align across the SV breakpoint site. On average, it is far more likely to encounter unbalanced split-reads than balanced split-reads, despite the later accurately predicting SV breakpoint regions. Both categories are used by RAPTR-SV to detect split-read SV sites.

References

1. Korbel JO, Urban AE, Affourtit JP et al (2007) Paired-end mapping reveals extensive structural variation in the human genome. Science 318:420–426. https://doi.org/10.1126/science.1149504
2. Chen K, Wallis JW, McLellan MD et al (2009) BreakDancer: an algorithm for high-resolution mapping of genomic structural variation. Nat Methods 6:677–681. https://doi.org/10.1038/nmeth.1363
3. Hormozdiari F, Hajirasouliha I, Dao P et al (2010) Next-generation VariationHunter: combinatorial algorithms for transposon insertion discovery. Bioinformatics 26:i350–i357. https://doi.org/10.1093/bioinformatics/btq216
4. Korbel J, Abyzov A, Mu X et al (2009) PEMer: a computational framework with simulation-based error models for inferring genomic structural variants from massive paired-end sequencing data. Genome Biol 10:R23. https://doi.org/10.1186/gb-2009-10-2-r23
5. Ye K, Schulz MH, Long Q et al (2009) Pindel: a pattern growth approach to detect break points of large deletions and medium sized insertions from paired-end short reads. Bioinformatics 25:2865–2871. https://doi.org/10.1093/bioinformatics/btp394
6. Handsaker RE, Korn JM, Nemesh J, McCarroll SA (2011) Discovery and genotyping of genome structural polymorphism by sequencing on a population scale. Nat Genet 43:269–276. https://doi.org/10.1038/ng.768
7. Layer RM, Chiang C, Quinlan AR, Hall IM (2014) LUMPY: a probabilistic framework for structural variant discovery. Genome Biol 15:R84. https://doi.org/10.1186/gb-2014-15-6-r84
8. Bickhart DM, Hutchison JL, Xu L et al (2015) RAPTR-SV: a hybrid method for the detection of structural variants. Bioinformatics 31:2084–2090. https://doi.org/10.1093/bioinformatics/btv086
9. Li H, Durbin R (2009) Fast and accurate short read alignment with Burrows-Wheeler transform. Bioinformatics 25:1754–1760. https://doi.org/10.1093/bioinformatics/btp324
10. Li H, Handsaker B, Wysoker A et al (2009) The sequence alignment/map format and SAMtools. Bioinformatics 25:2078–2079. https://doi.org/10.1093/bioinformatics/btp352
11. Hach F, Hormozdiari F, Alkan C et al (2010) mrsFAST: a cache-oblivious algorithm for short-read mapping. Nat Methods 7:576–577. https://doi.org/10.1038/nmeth0810-576

12. Hach F, Sarrafi I, Hormozdiari F et al (2014) mrsFAST-ultra: a compact, SNP-aware mapper for high performance sequencing applications. Nucleic Acids Res 42:W494–W500. https://doi.org/10.1093/nar/gku370
13. Zhang C-Z, Spektor A, Cornils H et al (2015) Chromothripsis from DNA damage in micronuclei. Nature 522:179–184. https://doi.org/10.1038/nature14493
14. Quinlan AR, Hall IM (2010) BEDTools: a flexible suite of utilities for comparing genomic features. Bioinformatics 26:841–842. https://doi.org/10.1093/bioinformatics/btq033
15. Hormozdiari F, Alkan C, Eichler EE, Sahinalp SC (2009) Combinatorial algorithms for structural variation detection in high-throughput sequenced genomes. Genome Res 19:1270–1278. https://doi.org/10.1101/gr.088633.108
16. English AC, Richards S, Han Y et al (2012) Mind the gap: upgrading genomes with Pacific Biosciences RS long-read sequencing technology. PLoS One 7:e47768. https://doi.org/10.1371/journal.pone.0047768

Chapter 12

Versatile Identification of Copy Number Variants with Canvas

Sergii Ivakhno and Eric Roller

Abstract

Versatile and efficient variant calling tools are needed to analyze large-scale sequencing datasets. In particular, identification of copy number changes remains a challenging task due to their complexity, susceptibility to sequencing biases, variation in coverage data and dependence on genome-wide sample properties, such as tumor polyploidy, polyclonality in cancer samples, or frequency of de novo variation in germline genomes of pedigrees. The frequent need of core sequencing facilities to process samples from both normal and tumor sources favors multipurpose variant calling tools with functionality to process these diverse sets within a single software framework. This not only simplifies the overall bioinformatics workflow but also streamlines maintenance by shortening the software update cycle and requiring only limited staff training. Here we introduce Canvas, a tool for identification of copy number changes from diverse sequencing experiments including whole-genome matched tumor–normal, small pedigree, and single-sample normal resequencing, as well as whole-exome matched and unmatched tumor–normal studies. In addition to variant calling, Canvas infers genome-wide parameters such as cancer ploidy, purity, and heterogeneity. It provides fast and easy-to-run workflows that can scale to thousands of samples and can be easily incorporated into variant calling pipelines.

Key words Copy number variation, Small pedigree, Somatic variation

1 Introduction

The increased throughput of sequencing studies has created high demand for versatile and scalable tools to detect somatic and germline copy number changes. Increasingly complex experimental designs require accurate characterization not only of individual copy number variants (CNVs) but also of global genome and sample properties, such as ploidy, normal contamination, and polyclonality (frequently present in cancer samples, [1]). The interaction of these factors creates an array of different somatic genome architectures that confounds optimization of CNV calling algorithms given an often limited availability of training data. Accurate variant and genotype calling is also crucial for successful identification of highly penetrant variants that cause rare disease, such as de novo or

Derek M. Bickhart (ed.), *Copy Number Variants: Methods and Protocols*, Methods in Molecular Biology, vol. 1833, https://doi.org/10.1007/978-1-4939-8666-8_12, © Springer Science+Business Media, LLC, part of Springer Nature 2018

recessive mutations [2]. Unfortunately, false positive and negative results can occur due to technical artifacts or reduced sequencing coverage, which especially impact CNVs identified through read depth estimation [3]. CNV calling accuracy in families can be improved over single-sample calling by incorporating pedigree structure into the genotyping model to ensure that copy number genotypes are consistent with Mendelian inheritance and low rates of de novo mutation.

While a number of methods for CNV identification have been introduced, most of them harbor shortcomings when it comes to scalability, throughput, and versatility. First, many CNVs callers target only very specific experimental designs, such as tumor–normal studies. In cases where handling of different samples and experimental set-ups is required, the need to manage multiple CNV calling tools creates significant overhead. Second, many tools require running additional third-party software to complete parts of the internal variant calling workflow (e.g., for segmentation), which complicates workflow management and version control. Third, model parameters to infer global genome and sample properties are often hard-coded and difficult to optimize for individual projects. Finally, while several tools have been developed for the identification of germline CNVs from sequencing data [4–6], most of them are limited to variant calling in single samples. Even those designed for family-based CNV detection are restricted to only deal with parent–offspring trios [6].

Canvas is designed to address the aforementioned limitations of existing solutions by offering the following features and functionality.

1. It fully implements all steps of the variant calling workflow and requires only aligned sequence data and related reference genome files as input.
2. Canvas offers inference of global tumor genome and sample characteristics, including ploidy, contamination and heterogeneity, as well as loss of heterozygosity.
3. Canvas is versatile, offering fast and easy-to-run whole-genome and exome workflows for both somatic and germline variant calling.
4. The small pedigree whole-genome workflow enables germline and de novo variant calling in pedigrees. In addition to trios, the small pedigree workflow can process quads and can also perform joint variant calling in batches of 10 samples.

This combined functionality makes Canvas a favorable tool for somatic and germline CNV detection in large-scale sequencing studies.

2 Materials

Canvas is implemented in the C# programming language and can be run on Linux and Windows systems using. NET Core (for versions 1.25.0 and above).

2.1 Installing. NET Core

On Linux download and extract the runtime tarball containing precompiled binaries for your operating system. On Windows, either download and run the runtime installer or download and extract the zip file containing precompiled binaries. For example, on Centos 7 run:

```
$ wget
https://download.microsoft.com/download/D/7/A/
D7A9E4E9-5D25-4F0C-B071-210CB8267943/dotnet-centos-
x64.1.1.2.tar.gz
$ mkdir dotnet-1.1.2
$ tar-xf dotnet-centos-x64.1.1.2.tar.gz -C
dotnet-1.1.2
$ dotnet-1.1.2/dotnet --version
```

2.2 Installing Canvas

Precompiled binaries are available from https://github.com/Illumina/canvas/releases. Simply download and extract the tarball and run them with. NET Core.

```
$ wget
https://github.com/Illumina/canvas/releases/
download/1.31.0.843%2Bmaster/Canvas-1.31.0.843.mas-
ter_x64.tar.gz
$ tar-xf Canvas-1.31.0.843.master_x64.tar.gz
$ dotnet-1.1.2/dotnet Canvas-
1.31.0.843+master_x64/Canvas.dll --help
```

3 Methods

Canvas supports a number of different workflows depending on the input sequencing data (Fig. 1). The available modes are:

- Germline-WGS: CaNV calling of a diploid germline sample from whole genome sequencing data.
- SmallPedigree-WGS: CNV calling of multiple diploid germline samples from whole genome sequencing data.
- Somatic-Enrichment: CNV calling of a somatic sample from targeted sequencing data.
- Somatic-WGS: CNV calling of a somatic sample from whole genome sequencing data.
- Tumor-normal-enrichment: CNV calling of a tumor–normal pair from targeted sequencing data.

3.1 Overview

The Canvas workflow comprises five distinct modules designed to (1) process aligned read data and calculate coverage bins, (2) per-

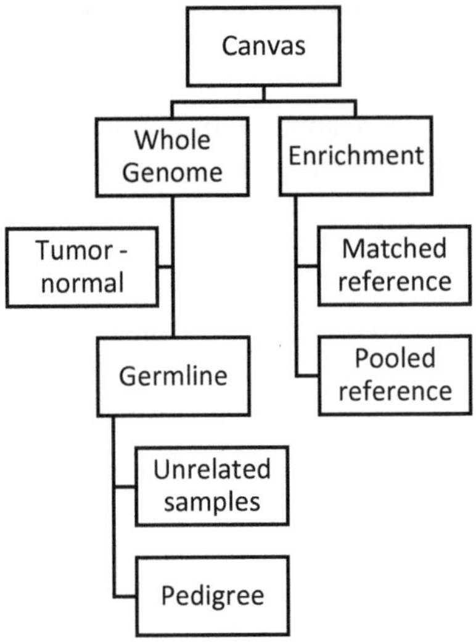

Fig. 1 Hierarchy of available Canvas workflows. Canvas enables analysis of both enrichment and whole-genome sequencing data across tumor and germline samples

form outlier removal and normalization of coverage estimates, (3) identify segments of uniform copy number, (4) calculate minor allele frequencies (MAF) and (5) assign segment copy numbers and infer genome-wide properties. Separate workflows exist for somatic, germline (including small pedigree), and exome sequencing data. The latter workflow can be run either with or without a matched normal control sample and requires a manifest file with targeted regions. For brevity, below we highlight core functionalities of individual modules; an example diagram for the small pedigree workflow is provided in Fig. 2. We also describe output files produced by individual modules. Algorithmic details of individual components are available from Roller et al. [7], while the pedigree variant calling algorithm is described in Ivakhno et al. [8].

Most users will not need to access individual modules, as the main Canvas executable conveniently manages all workflow tasks. However, the knowledge of individual modules might be useful for debugging and for providing optional parameters to the components.

3.2 Coverage Binning (CanvasBin)

Genomic bins are determined such that the number of unique k-mers within each bin is fixed. This leads to a variable bin size along the genome that depends on repeat content and alignment complexity. CanvasBin iterates through the input bam file and calculates the number of alignments falling within each bin. The output of CanvasBin is a gzipped BED-like file (.binned) containing chromosome name, genomic start position, genomic stop position, observed alignments and GC content for each bin.

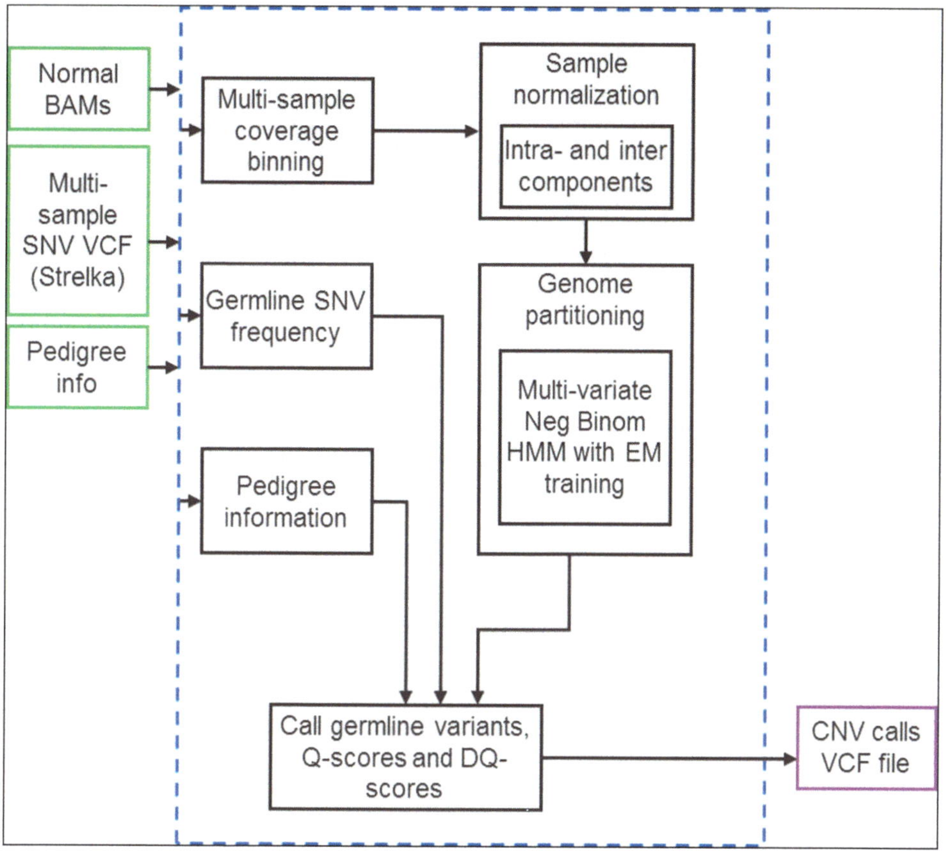

Fig. 2 Schematics of the Canvas Small Pedigree Workflow inputs, outputs, and stages. Three main components are outlined: data preprocessing and normalization, multisample segmentation, and joint de novo and inherited germline CNV calling

3.3 Normalization (CanvasClean)

CanvasClean implements four procedures to clean up the coverage signal from CanvasBin:

1. Bin coverage outlier removal.
2. Bin window size outlier removal.
3. GC content correction.
4. Normalization of formalin-fixed, paraffin-embedded (FFPE) samples (somatic workflows only).

Procedure 4 (FFPE Normalization) can optionally include both bin-level denoising and fragment-based GC-content normalization. CanvasClean outputs a. cleaned file with the same format as the. binned file produced by CanvasBin.

3.4 Single-Sample Partitioning (CanvasPartition) and MAF Calculation (CanvasSNV)

For single-sample analysis, CanvasPartition implements a coverage partitioning algorithm based on the unbalanced Haar wavelet transform [9]. CanvasPartition outputs a gzipped BED-like file (.partitioned) containing chromosome name, genomic start position, genomic stop position, bin coverage and an integer label that identifies which partitioned segment the bin belongs to.

CanvasSNV computes the MAF at heterozygous SNV positions. These candidate SNV positions are provided by a VCF file containing either germline SNVs called in each sample (or matched normal sample for somatic workflows) or SNV sites common in the population (such as from the 1000 Genomes Project or dbSNP). The output of CanvasSNV is a gzipped tab-delimited text file containing one header line (beginning with #) and one line for each heterozygous SNV. Each SNV line contains the chromosome name, position, reference allele, alternate allele, number of reads supporting the reference allele, and number of reads supporting the alternate allele.

3.5 Multisample Segmentation (CanvasPartition)

When analyzing multiple samples, CanvasPartition considers the coverage from all samples for each bin. A data matrix of coverage values is produced where each row represents the coverage of all samples for a given genomic bin. In turn, each column represents the normalized coverage across all genomic bins for a given sample. Partitioning of this data matrix is achieved through a hidden Markov model (HMM) with multivariate negative binomial emission distribution, where hidden states are initialized to approximately follow copy number states (exact copy number assignment is done at the variant calling stage). First, the Expectation Maximization algorithm is used to optimize parameters of the emission and transition distributions. Next, the Viterbi algorithm is used to derive the final partitions.

3.6 Somatic Model and Copy Number (CN) Assignment for Tumor–Normal Whole-Genome and Exome Workflows (CanvasSomaticCaller)

Accurate somatic copy number assignment requires knowledge of genome ploidy, sample purity (normal contamination level) and location of heterogeneous variants. To infer these parameters CanvasSomaticCaller fits a deviation model based on the principle that for a given combination of purity and diploid coverage level, the expected MAF and coverage values for each CN state can be easily calculated. CanvasSomaticCaller calculates a model deviation D for each combination in a range of purities and diploid coverage levels by comparing observed and expected MAF and coverage values. To better discriminate between models with similar deviation values, a separate penalty term P is introduced to measure the complexity of copy number variation implied by each model. Variables that best predict genome complexity are inferred using logistic regression on a training set of sequenced tumor genomes with known karyotypes. Finally, a separate EM clustering module is used to group segments into clusters of similar MAF and coverage and assess if they could represent heterogeneous variants. This is done by computing the distance between observed cluster centroids and expected centroids if there were no heterogeneity. The resulting cluster deviation C is combined with the other two terms to provide a total deviation $T = D$ *(model deviation)* $+ P$ *(penalty term)* $+ C$. Ploidy, purity, and heterogeneity values from the model with the smallest total deviation are then used to assign a copy number genotype to each segment.

The primary output from CanvasSomaticCaller is a bgzipped VCF file. The VCF file header includes the estimated tumor purity (1.0 for pure tumor), the estimated overall ploidy (mean copy number across all autosome bases; 2.0 for the reference genome), tumor heterogeneity proportion and the estimated chromosome count (total number of chromosomes weighted by their copy number).

3.7 Variant Calling for Germline Variants (Canvas-PedigreeCaller)

For each segment identified by the multisample CanvasPartition module, CanvasPedigreeCaller uses the distribution of coverage and MAF to assign copy numbers. As a rule, allele-specific copy numbers are determined only when a segment contains enough heterozygous SNVs; small segments containing only a handful of SNV sites will be assigned a total copy number, but not allele-specific copy numbers. A probabilistic model is used to estimate the maximum likelihood (L) of copy number (CN) assignments within a trio (M-mother, F-Father, C-Child) given the observed coverage data (D)

$$L\left(CN_{M'}CN_F,CN_C / D\right) \sim P\left(D_M / CN_M\right)P\left(D_F / CN_F\right)$$
$$P\left(D_C / CN_C\right)\times P\left(CN_C / CN_M,CN_F\right)$$

where the last term incorporates both the Mendelian transmission probabilities and the probabilities of de novo events. Likelihood evaluation is done using exhaustive enumeration of all possible CN assignments within the pedigree up to the maximum CN threshold. The table of joint probabilities corresponding to the model with the maximum likelihood is used to estimate a quality score for each copy number genotype call. For each copy number genotype call that is inconsistent with Mendelian inheritance a de novo quality score is assigned.

The primary output from CanvasPedigreeCaller is a multisample VCF file compliant with the VCF Version 4.1 specification. The copy number (CN) and major chromosome count (MCC) for each segment are indicated in the per-sample genotype field along with the corresponding quality score (QS). De novo calls also have a Phred-scaled de novo quality (DQ) score the same field. De novo events in a child sample with DQ score greater than 20 can be viewed using bcftools:

```
$ bcftools view -s child CanvasOutdir/CNV.vcf.gz |
bcftools filter -i 'FORMAT/DQ>20'
```

The FILTER column in the multisample VCF file will be PASS if any sample has a passing variant call as indicated by the per-sample filter tag (FT). In addition to the VCF file, a coverage.bedgraph file (and coverage.bigWig file on Linux) of coverage is produced for each sample with sample ID {SampleID} in the directory TempCNV_{SampleID} under the output directory. These files contain the normalized coverage values for each bin produced by CanvasBin. Normalized coverage values correspond to ploidy, such that a diploid region will have coverage 2.0, a triploid region will have coverage 3.0 and so on. These files can be loaded into IGV or another compatible genome browser.

3.8 Running Canvas

Canvas can be run on a variety of sequencing inputs. As exact command line parameters may vary between software versions, the best way to obtain the relevant command line options is to run Canvas.dll with the help option (-h). To get the list of available workflows (*see* Subheading 3) run Canvas.dll with just the help option. To get the list of available parameters for a particular workflow run Canvas.dll with the specified workflow name followed by the help option:

```
$ dotnet Canvas.dll SmallPedigree-WGS -h
```

The help information will indicate which parameters are required and which parameters are optional along with their default values if any.

As exemplified in the previous sections, Canvas comprises different executable components wrapped into easy-to-use workflows for different sample types and sequencing assays. While the workflows can be invoked through a single command line to generate a set of copy number calls without any further intervention, users can also run individual components to generate intermediate files or rerun individual components with different parameters. For example, running "dotnet CanvasBin.dll" will list all input arguments to the coverage binning module. Modification of individual component parameters can be achieved by passing them to the main Canvas.dll executable using the custom parameters option:

```
--custom-parameters [module_name],[parameter_ name]=
[parameter_value]
```

For example, to enable fragment-based GC-content normalization beneficial for processing FFPE samples through the somatic workflows, the CanvasBin "mode" parameter ("-m") can be altered through Canvas.dll command line argument:

```
--custom-parameters=CanvasBin,-m=TruncatedDynamicRange
```

The custom parameters option can be provided on the Canvas.dll command line multiple times to configure other options for the same module or any other modules.

In addition to running individual Canvas modules, Canvas allows running a subset of modules by utilizing checkpoints through the following options:

- -c = VALUE continue analysis starting at the specified checkpoint. VALUE can be the checkpoint name or number
- -s = VALUE stop analysis after the specified checkpoint is complete. VALUE can be the checkpoint name or number

Canvas modules have the following checkpoint numbers: 2 for Canvas Clean, 3 for CanvasBin, 4 for CanvasClean, 5 for CanvasPartition, and 6 for variant calling although they are subject to change in different versions of Canvas. Omitting -s will execute Canvas until the final step in the workflow has completed. For example, specifying -c 4 in the whole-genome somatic workflow will run CanvasClean, CanvasPartition, and CanvasSomaticCaller. This could be advantageous if users want to rerun Canvas with different a partitioning algorithm without the need to repeat the time consuming I/O steps of CanvasBin and CanvasSNV. Ease of module reuse is further expanded by the fact that Canvas intermediate files are saved as BED-like files, facilitating access by third-party software such as bedtools.

3.9 Canvas Annotation Files

The required input files for Human reference genome builds GRCh37, hg19, and GRCh38 can be downloaded from S3 at http://canvas-cnv-public.s3.amazonaws.com. Any browser can be used to download these files or by using the command line tool wget on Linux systems. For example, to download the hg19 GenomeSize.xml file run.

```
wget http://canvas-cnv-public.s3.amazonaws.com/
hg19/WholeGenomeFasta/GenomeSize.xml.
```

When using a custom reference genome the equivalent files that are available for download will need to be created. Use the FlagUniqueKmers utility shipped with each Canvas release to generate the annotated fasta file (kmer.fa) for a custom reference genome. The FlagUniqueKmers utility is available in the Tools subdirectory of every release tarball and also runs under. NET Core (*see* Subheading 2.2).

3.10 Working Example Using the Canvas Small Pedigree Workflow

Here we focus on the Small Pedigree Workflow to provide a step-by-step guide to Canvas usage. Whole genome sequencing is becoming a standard approach in clinical genetics for identifying highly penetrant de novo and recessive variants that cause rare diseases. Small pedigree sequencing is a powerful technique for identifying de novo CNVs, which makes it an ideal use case for this demo. Since the de novo rate for large variants in pedigrees is very low (less than 0.1 variants per offspring per generation, [2]), we selected a synthetically simulated trio (bam files of 60x coverage) for this example. The synthetic dataset contains a higher rate of de novo variants and fully demonstrates the ability of Canvas to detect CNVs with a broad range of sizes. Additionally, the synthetic dataset provides a comprehensive truth set to assess the accuracy of Canvas (the availability of comprehensive truth sets for biological samples is limited, especially for a full trio, as curation can be a slow, resource-intensive and error-prone process). We used the EvaluateCNV tool (bundled with each Canvas release tarball) to assess two key performance characteristics of the Small Pedigree Workflow: (1) ability to accurately call inherited germline variants and (2) correct de novo variant detection. Amazon AWS m4.4xlarge Ubuntu 16.04 instance was used to create this demo running Canvas version 1.31.0.843. An alternative instance type or local cluster node could also be used, though it is recommended that the amount of RAM per core is 4 GB.

3.11 Data and Binaries

Input samples used in this demo represent a synthetically simulated trio generated by adapting a haplotype subsampling strategy described in Ivakhno et al. [10]. Briefly, phased SNV and indel variants for NA12882 and NA12878 from the Platinum Genomes pedigree [11] have been used to split aligned reads into separate haplotypes, one for each chromosome. Next, the gHapMix algorithm was used to simulate CNV variants in a pedigree from resulting haploid bams leading to an average of 7500 variants per trio and 200 de novo variants in a proband. The steps below will enable access to example simulation data. More details of simulation strategy can be found in Ivakhno et al. [8], Supplementaty Material Subheading 3.1, **step 2**.

1. Canvas version 1.31.0.843 binary was compiled using. Net Core 1.1.2, hence we will install it on our demo using applications installer (users can also follow 2.1 instruction to download the binary directly)

```
sudo sh -c 'echo "deb [arch=amd64] https://apt-mo.trafficmanager.net/repos/dotnet-release/ xenial main" > /etc/apt/sources.list.d/dotnetdev.list'
sudo apt-key adv --keyserver hkp://keyserver.ubuntu.com:80 --recv-keys 417A0893
sudo apt-get update
sudo apt-get install dotnet-dev-1.1.0
```

2. Download Canvas binary (${CanvasDIR} in this demo) and register for a BaseSpace account https://basespace.illumina.com/.
3. Add BaseSpace project https://basespace.illumina.com/s/flganFhSPsBo with simulation bams to your account.
4. Install BaseMount and load the canvas-spw project

```
sudo bash -c "$(curl -L https://basemount.basespace.illumina.com/install )"
mkdir /tmp/BaseSpace
basemount --scopes="Create Global, Browse Global, Create Projects, Read Global" /tmp/BaseSpace
```

5. This should show the following folders under
   - bams = simulated trio bams of 60x coverage aligned with Isaac

- canvasdata = hg19 genome reference files for running Canvas (can also be downloaded from S3 http://canvas-cnv-public.s3.amazonaws.com)
- snvvcf = SNV vcf files to accompany bams (joint germline CNV calls using Strelka2 https://github.com/Illumina/strelka)
- simdata = bed files with simulated inherited and de novo variants

3.12 Executing Canvas

In this example we are accessing files through basemount (Canvas should be run as a user rather than sudo root). Files could also be copied to a local drive and run from there. Small variants used in this example (in Pedigree.vcf.gz file) were derived by using Strelka2 variant caller [12], which is recommended for use with Canvas.

Issue the following command (workdir=/tmp/BaseSpace/Projects/canvas-spw/AppResults):

```
dotnet $CanvasDIR/Canvas.dll SmallPedigree-WGS
--bam=$workdir/bams/Files/father.bam father
--bam=$workdir/bams/Files/mother.bam mother
--bam=$workdir/bams/Files/child1.bam proband
-r $workdir/canvasdata/Files/kmer.fa -g $workdir/
canvasdata/Files/* --sample-b-allele-vcf $workdir/
snvvcf/Files/Pedigree.vcf.gz -f $workdir/canvasdata/
Files/filter13.bed -o /tmp/gHapMixDemo
```

*Note that some versions of Canvas require genome reference (-g) to have a particular sub-path. If there is an error, copy the $workdir/canvasdata/Files/GenomeSize.xml file into a new directory with the sub-path specified in the error message.

The runtime will depend on the number of available CPUs and whether bam files were copied to a local drive. The run on a bare Amazon m4.4xlarge instance (16 CPUs and 64G RAM) with network I/O took 3.5 h. Results are available as VCF files: either a multisample VCF under gHapMixDemo or single-sample equivalents under gHapMixDemo/TempCNV folders.

3.13 Assessing Accuracy with EvaluateCNV

Here we will use the EvaluateCNV tool supplied with Canvas distribution to calculate various performance metrics for inherited and de novo CNVs. Users should note that different versions of Canvas might produce slightly varied results.

First, we can run EvaluateCNV to produce recall and precision metrics for inherited Canvas CNV calls using truth variant files.

```
touch generic.cnaqc.excluded_regions.bed
(EvaluateCNV can exclude regions from evaluation,
but here we won't be using this)
dotnet $CanvasDIR/Tools/EvaluateCNV/EvaluateCNV.dll
/tmp/BaseSpace/Projects/CanvasSPW/AppResults/
simdata/Files/child1_truth.bed /tmp/gHapMixDemo/
```

```
TempCNV_child1/CNV.vcf.gz    generic.cnaqc.excluded_
regions.bed /tmp/gHapMixDemo/TempCNV_child1/
```

Results will be written to /tmp/gHapMixDemo/TempCNV_child1/EvaluateCNVResults.txt

This gives us for PASS inherited variants (in inheritedCNVs.txt) recall of 90.08 and precision of 94.41%.

Next, we run the same command but using the de novo variant truth file and a -q 20 argument to extract variants with de novo quality score above 20 (default for Canvas).

```
dotnet $CanvasDIR/Tools/EvaluateCNV/EvaluateCNV.dll
/tmp/BaseSpace/Projects/CanvasSPW/AppResults/
simdata/Files/child1_truth_denovo.bed          /tmp/
gHapMixDemo/TempCNV_child1/CNV.vcf.gz      generic.
cnaqc.excluded_regions.bed         /tmp/gHapMixDemo/
TempCNV_child1/ -q 20
```

Results will be written to /tmp/gHapMixDemo/TempCNV_child1/EvaluateCNVResults_denovo.txt

This gives us for PASS de novo variants (in denovoCNVs.txt) recall of 100.00% and precision of 89.91%.

3.14 Visualization

Visualization of calls in IGV or other genome browsers allows for validation of CNV calls. Figure 3 lists an example custom visualization of homozygous de novo deletion at coordinates chr1:45,485,881–45,531,192 using the. partitioned file. In addition, examples of population coverage and CNV variant tracks are provided—these can be used to assess false positive calls, i.e., when a de novo variant is identical to a common variant in populations.

3.15 Further Documentation

In this chapter we provided an introduction to the various Canvas workflows, described key characteristics of individual Canvas modules and provided a demonstration of the Canvas small pedigree workflow. For an in-depth discussion of Canvas internal algorithms, benchmarking, and comparison with third-party tools readers should consult supplementary material to Roller et al. [7] and Ivakhno et al. [8]. Ivakhno et al. [8] specifically covers Canvas applications to variant calling in pedigrees. As actively developed software, Canvas versions with new functionality and bug fixes are released frequently. Readers are encouraged to consult https://git.illumina.com/Bioinformatics/Canvas for the latest release and documentation or to ask questions, submit issues and contribute to development.

3.16 Running Canvas as an Illumina® Sequencing Analysis App

In this chapter we focused on describing the standalone Canvas workflow which can be deployed on either local clusters or cloud-based systems. Using a standalone deployment allows greater customization to cater to different user needs. For example, Canvas

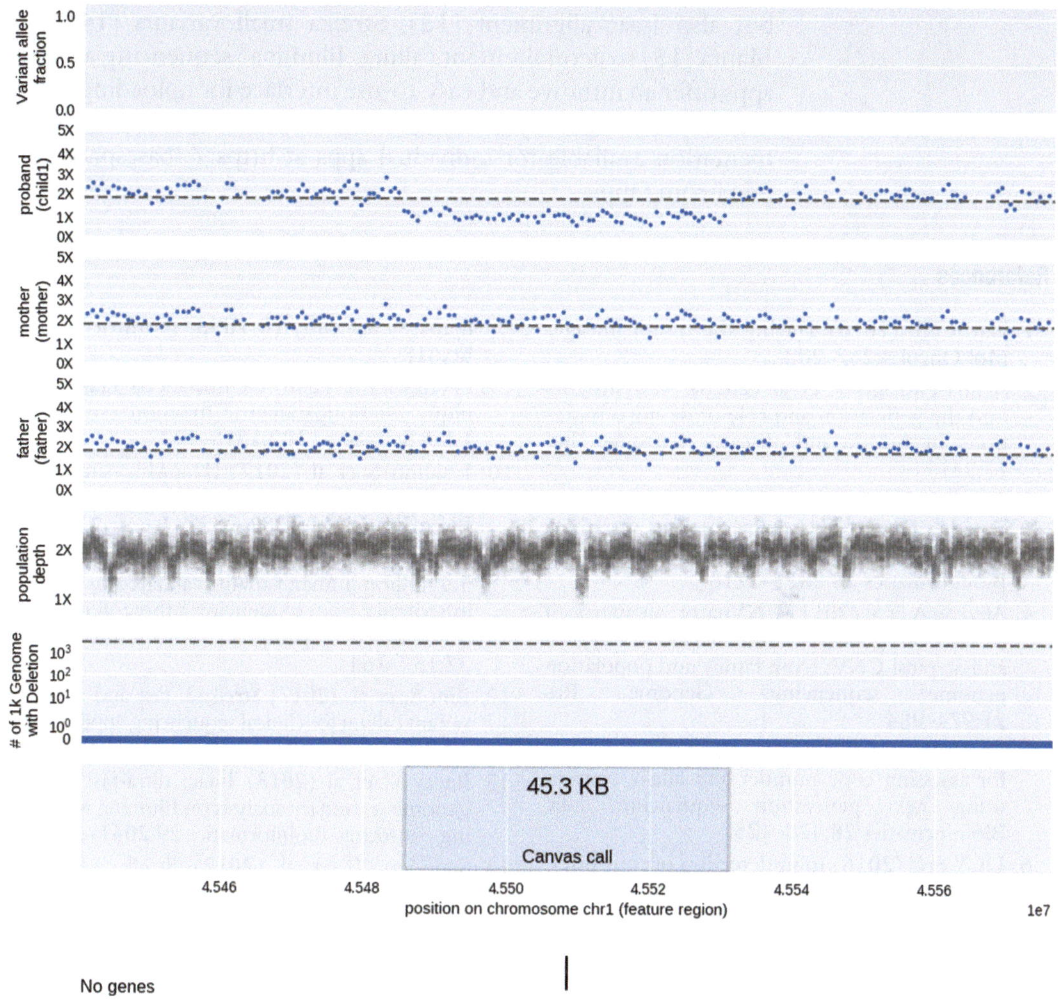

Fig. 3 Example visualization of a Canvas de novo call in a proband (child1). The call clearly overlaps the simulated heterozygous deletion identifiable by low coverage in the proband and diploid coverage in the parents. Preannotated population variants such as the ones derived from the 1000 Genomes Project can be used as an extra filter when inspecting de novo calls for false positive signal

could be integrated into existing sequence analysis workflows and run on local clusters administered by core facilities. This could be a cost-effective solution for institutes with high sequencing throughput. However, infrequent users without local bioinformatics support might find it difficult to configure and install Canvas themselves. An alternative way of running Canvas is to use one of the Illumina sequencing analysis apps available through BaseSpace™ Sequence Hub. Currently available workflows include "Whole Genome Sequencing", "Tumor Normal" and "Small Pedigree" apps. These apps will provide not only CNV calling from Canvas, but also Isaac alignment [13], Strelka small variants [14] and Manta [15] structural variant calling. Illumina® sequencing analysis apps offer an intuitive and easy-to-use interface for uploading input files and running analyses. For details readers can consult the documentation available for individual apps at https://basespace.illumina.com/apps.

References

1. Navin NE (2010) Tracing the tumor lineage. Mol Oncol 4:267–283
2. Acuna-Hidalgo R et al (2016) New insights into the generation and role of de novo mutations in health and disease. Genome Biol 17:241
3. Teo SM et al (2012) Statistical challenges associated with detecting copy number variations with next-generation sequencing. Bioinformatics 28:2711–2718
4. Abyzov A et al (2011) CNVnator: an approach to discover, genotype, and characterize typical and atypical CNVs from family and population genome sequencing. Genome Res 21:974–984
5. Boeva V et al (2011) Control-FREEC: a tool for assessing copy number and allelic content using next generation sequencing data. Bioinformatics 28:423–425
6. Liu Y et al (2016) Joint detection of copy number variations in parent-offspring trios. Bioinformatics 32:1130–1137
7. Roller E et al (2016) Canvas: versatile and scalable detection of copy number variants. Bioinformatics 32:2375–2377
8. Ivakhno S, et al. (2018) Canvas SPW: calling de novo copy number variants in pedigrees. https://doi.org/10.1093/bioinformatics/btx618
9. Fryzlewicz P (2007) Unbalanced Haar technique for nonparametric function estimation. J Am Stat Assoc 102:1318–1327
10. Ivakhno S et al (2017) tHapMix: simulating tumour samples through haplotype mixtures. Bioinformatics 33:280–282
11. Eberle M et al (2016) A reference dataset of 5.4 million human variants validated by genetic inheritance from sequencing a three-generation 17-member pedigree. Genome Res 27:157–164
12. Kim S, et al. (2017) Strelka2: fast and accurate variant calling for clinical sequencing applications. bioRxiv. https://doi.org/10.1101/192872
13. Raczy C et al (2013) Isaac: ultra-fast whole-genome secondary analysis on Illumina sequencing platforms. Bioinformatics 29:2041–2043
14. Saunders C et al (2012) Strelka: accurate somatic small-variant calling from sequenced tumor–normal sample pairs. Bioinformatics 28:1811–1817
15. Chen X et al (2016) Manta: rapid detection of structural variants and indels for germline and cancer sequencing applications. Bioinformatics 32:1220–1222

Chapter 13

A Randomized Iterative Approach for SV Discovery with SVelter

Xuefang Zhao

Abstract

Genomic structural variants (SVs) are major sources of genome diversity, and numerous studies over the past few decades have shown the impact this class of genetic variation has had on human health and disease. In spite of the recent advances in sequencing technology and discovery methodology, there are still considerable amount of variants in the genome that are partially or completely misinterpreted. The computational tool introduced in this chapter, SVelter, is specifically designed to detect and resolve genomic SVs in all different formats, including the canonical as well as the complex.

Key words Structural variation (SV), Next generation sequencing, Randomized iterative process, Copy number variant (CNV)

1 Introduction

Structural variants (SVs), defined as the removal or rearrangement of large DNA pieces, are natural sources of genetic variation [1, 2] that have been implicated in numerous human diseases [3–5]. There have been extensive studies to identify these genomic aberrations within the whole genomes of humans and other species and numerous algorithms have been developed to accurately identify their prevalence [6–10].However, most of these approaches have primarily focused on canonical forms of SVs such as copy number variants (CNVs; deletions, duplications) or copy neutral (inversions) rearrangements defined by at most two chromosomal breakpoints (BPs) [11]. While such SVs are common in the genome and as a class represent the majority of SVs previously reported, there are additional forms of rearrangements that are far more convoluted. These complex SVs (CSVs) typically involve three or more BPs and result from multistep or accumulative genomic rearrangements, of which the potential role in pathogenesis of cancers and various neurological developmental disorders has been indicated in numerous studies [3, 4]. However, CSVs are usually misinterpreted

Derek M. Bickhart (ed.), *Copy Number Variants: Methods and Protocols*, Methods in Molecular Biology, vol. 1833, https://doi.org/10.1007/978-1-4939-8666-8_13,

by most currently available algorithms due to misleading alignment signals presented in such cases.

The software introduced in this chapter, SVelter [12], is capable of detecting SVs in both canonical and complex forms. Unlike traditional methods that infer the SVs by analyzing the deviant alignment signals, SVelter implements a Markov chain Monte Carlo (MCMC) procedure whereby genomic segments are virtually rearranged in a randomized fashion to reach the state with minimized aberrations relative to the observed characteristics of the sequence data. In this manner, SVelter is able to interrogate many different types of rearrangements, including multideletion and duplication–inversion–deletion events as well as distinct overlapping variants on homologous chromosomes. The framework is provided as a publicly available software package that is available online (https://github.com/mills-lab/svelter).

The primary purpose of this chapter is to demonstrate the proper protocol for applying SVelter and will guide users from the alignments of paired end whole genome sequences (WGS) to the final SV predictions indexed in VCF format. Moreover, specific techniques implemented in the software that help minimize the overall runtime will also be discussed in detail so that users can balance between their preferred runtime and computing resources.

2 Materials

Software Preparation:

For SVelter to be properly applied, R (*https://www.r-project.org/*), Python (*https://www.python.org/*) and samtools (*http://samtools.sourceforge.net/*) are required to be downloaded and properly installed. For R, the "*mixtools*" package is required (*https://cran.r-project.org/web/packages/mixtools/index.html*). For python, packages such as *numpy, scipy, random, itertools*, and *cython* are also necessary.

SVelter is an open source tool and can be easily obtained from the Mills Lab GitHub page with the following command:

```
git clone https://github.com/mills-lab/svel-
ter.git
cd svelter/
python setup.py install
```

3 Methods

Data Preparation:

Sample NA12878 has been deep-sequenced by various different projects including The 1000 Genomes Project (1KGP) and Genome in a Bottle (GIAB), with SVs across the whole genome systematically discovered and presented in numerous publications [13, 14]. The high coverage alignment of this sample produced by the 1000 Genomes Project, together with the corresponding reference genome, were obtained and processed by SVelter for whole spectrum SV discovery as an exemplar for its application. Alignment in both BAM and CRAM formats are acceptable by SVelter (*see* **Note 1**); CRAM is downloaded as an example here with the following linux command:

```
wget
ftp://ftp.1000genomes.ebi.ac.uk/vol1/ftp/
phase3/data/NA12878/high_coverage_alignment/
NA12878.mapped.ILLUMINA.bwa.CEU.high_cover-
age_pcr_free.20130906.bam.cram
wget
ftp://ftp.1000genomes.ebi.ac.uk/vol1/ftp/
phase3/data/NA12878/high_coverage_alignment/
NA12878.mapped.ILLUMINA.bwa.CEU.high_cover-
age_pcr_free.20130906.bam.cram.crai
```

download and index reference genome:

```
wget
ftp://ftp.1000genomes.ebi.ac.uk/vol1/ftp/
technical/reference/human_g1k_v37.fasta.gz
gunzip human_g1k_v37.fasta.gz
samtools faidx human_g1k_v37.fasta
```

3.1 Setup Working Environment

It is important to properly setup the working environment for SVelter before applying the tool, which can be easily accomplished by running its 'Setup' module in SVelter. Assuming *../workdir/* as the writable working directory, *../workdir/reference/* as the folder containing the reference genome, *../workdir/sample/* as the folder for sample sequences, and the software is kept under */local/lib/svelter/,* the module can be applied as follows (*see* **Note 2**):

```
svelter.py Setup --reference ../workdir/ref-
erence/human_g1k_v37.fasta
--workdir ../workdir/ --support /local/lib/
svelter/Support/GRCh37/
```

3.2 Estimate Default Distribution of Multiple Parameters

The existence of genomic SVs is usually indicated by aberrant alignment signatures such as abnormally long / short insert sizes, discordant pair orientations and clipped reads referred to as reads that are partially aligned to the reference genome. The definition of "aberrance" is relative to the "normal" which is approximated by randomly sampling copy neutral genomic regions. The insert size library, overlap read depth distribution were fitted with carefully designed statistical models, the details of which were explained in [12]. These models can be generated empirically by SVelter as follows:

```
svelter.py NullModel -workdir ../workdir/
--sample ../workdir/sample/NA12878.cram
```

3.3 Search for Chromosomal Breakpoints

Chromosomal breakpoints (BPs) define the edges of genomic structural rearrangements and thus it is important to first have precise BP candidates before SVs can be accurately identified and resolved. During the "BPSearch" module, SVelter traverses the genome to cluster pairs with aberrant alignment signals such as abnormal insert sizes or inconsistent pair orientations and assigns breakpoints accordingly. For larger multichromosomal genomes, e.g., human, the BPSearch step enables each chromosome to be processed independently which allows the user to run multiple jobs in parallel to minimize the runtime of the tool. As an example, chromosome22 in NA12878 can be queued as follows:

```
svelter.py BPSearch --workdir ../workdir/
--sample ../workdir/sample/NA12878.cram
--chromosome 22
```

3.4 Cluster Breakpoints

The "BPIntegrate" module clusters related breakpoints that are potentially involved in the same genomic rearrangements and prepare the breakpoint clusters in text files for the downstream analysis. SVelter allows multiple jobs to run in parallel to minimize the runtime, which is accomplished by setting different parameters in this step: cluster breakpoints genome wide, per chromosome or by length of defined size.

To cluster breakpoints genome wide:

```
svelter.py BPIntegrate --workdir ../workdir/
--sample ../workdir/sample/NA12878.cram
```

To prepare breakpoint clusters on each chromosome separately, simply add "--batch 0" to the command; to produce the output in predefined size, add "--batch X" instead where X stands for the desired number of clusters in each file.

3.5 Resolve Genomic SVs Through the MCMC Procedure

Breakpoint clusters from the previous step are integrated in txt format under *../workdir/ bp_files.sample_file/*, and each can be processed independently in this step.

```
svelter.py SVPredict --workdir ../work-
dir/ --sample ../workdir/sample/NA12878.
```

```
cram --bp-file /workdir/bp_files.NA12878.cram/
NA1278.file_name.bam.txt
```

3.6 Integrate the SVs into VCF Format

This is the final step of SVelter, during which the structural genomic variants discovered in previous steps are integrated into an easily-interpretable standardized format. Users can specify the output file through the "—prefix" parameter. In this example, output files "output_name.vcf" and "output_name.svelter" are exported from this step.

```
svelter.py SVIntegrate -workdir ../workdir/
--prefix output_name -input-path
../workdir/bp_files.Sample.bam/
```

3.7 Output Format

The output from SVelter is reported in two equivalent formats: *output_name.vcf* and *output_name.svelter*.

3.7.1 VCF Format

SVelter categorizes SVs into discrete classes, including *DEL*, *DUP*, and *INV* for simple SVs and *DEL_INV*, *DUP_INV*, *DEL_DUP*, *DEL_DUP_INV* and *others* for more complex rearrangements.

Canonical SVs are encoded in the same format with standard VCF4.1, while complex SVs have their unique formats, with examples of each category displayed in Table 1. *DEL_INV* refers to the events where a deletion locates next to an inversion, and these events are usually misinterpreted as overlap inversions through the traditional paired end approaches. In the VCF format, deleted and inverted blocks are specified in the info column with "del=" and "inv=" (*see* Table 2).

DUP_INVs differentiate themselves from the canonical duplications with the fact that the extra copies in the sample genome are in the opposite orientation compared against the original. Genomic coordinates of the original genomic piece is specified in the first two columns as well as "END=" in the info column. Insertion point of the duplicated piece is specified as "insert_point=" (*see* Table 3). DEL_DUP_INVs are similar to DUP_INVs, but differs with a deletion usually at where the duplicated region is inserted. Coordinates of the deleted region is specified by "del=" in the info column (*see* Table 4).

DEL_DUPs are the cases where a deletion locates next to a duplication, both are specified in the info column with "del=" and "dup=" respectively.

3.7.2 SVelter Format

SVelter format is a tab-delimited file with several additional columns that provide extended information (*see* Table 5). The first three columns specify the start and end positions of the event. The "bp_info" column explicitly lists all the breakpoints involved in the

Table 1
Example of DEL_INV, where a deletion locates next to an inversion, in VCF format

#CHR	POS	ID	REF	ALT	QUAL	FILTER	INFO	FORMAT	NA12878.hg19.chr1.vcf
chr1	187,464,829	chr1: 187,464,829:187465015: 187466725	c	<DEL_INV>	87	PASS	SVTYPE = del_inv;END = 187,466,725;del = chr1: 187464829–187,465,015;inv. = chr1:187465015–187,466,725; other = ab/ab_ab/b^_chr1:187464829:187465015:187466725	GT	0/1

Table 2
Example of DUP_INV, in VCF format

#CHR	POS	ID	REF	ALT	QUAL	FILTER	INFO	FORMAT	NA12878.hg19.chr1.vcf
chr1	81,661,376	chr1: 81,660,357:81661376: 81661553	a	<DUP_INV>	83	PASS	SVTYPE = dup_inv;END = 81,661,553;insert_point = chr1: 81660357;other = ab/ab_ab/b^ab_chr1:81660357: 81661376:81661553	GT	0/1

Table 3
Example of DEL_DUP_INV, in VCF format

#CHR	POS	ID	REF	ALT	QUAL	FILTER	INFO	FORMAT	NA12878.hg19.chr1.vcf
chr1	192,799,455	chr1: 192,799,455:192991952:192992496: 193012339:193043099	c	<DEL_DUP_ INV>	91	PASS	SVTYPE = del_dup_inv;END = 192,991,952; del = chr1:192991952–192,992,496; del = chr1:193012339–193,043,099; dup_inv = chr1:192799455–192,991,952; insert_point = chr1:193012339; other = abcd/abcd_aca^/db_chr1: 192799455:192991952:192992496: 193012339:193043099	GT	0/1

Table 4
Example of DEL_DUP, in VCF format

POS	ID	REF	ALT	QUAL	FILTER	INFO	FORMAT	NA12878.hg19.chr1.vcf
11,101,600	chr1: 11,101,600:11101804: 11101966:11102029	g	<DEL_DUP>	94	PASS	SVTYPE = del_dup;END = 11,101,966; del = chr1:11101600–11,101,804; dup = chr1:11101804–11,101,966; other = abc/abc_abc/bbc_chr1:11101600: 11101804:11101966:11102029	GT	0/1

Table 5
SVelter output file format

chr	start	end	bp_info	ref	alt	sv_class	geno-type	other_info
chr1	869,425	870,270	chr1:869425:870270	a/a	/	del	1/1	
chr2	125052911	125053244	chr2:125051757:125052911:125053244	ab/ab	ab/b^ab	dup_inv	0/1	insert_point=chr2:125051757
chr2	4205204	4205581	chr2:4205204:4205581:4212686:4223509	abc/abc	ab/aba^	del_dup_inv	0/1	del=chr2:4212686-4223509; dup_inv=chr2:4205204-4205581; insert_point=chr2:4212686
chr2	184571064	184572011	chr2:184569179:184569768:184571064:18 4572011	abc/abc	abc/c^bc	del_dup_inv	0/1	del=chr2:184569179-184569768; dup_inv=chr2:184571064-184572011; insert_point=chr2:184569179
chr3	57387037	57387317	chr3:57386892:57387037:57387162:57387 317	abc/abc	abc/acc	del_dup	0/1	del=chr3:57387037-57387162; dup=chr3:57387162-57387317

Table 6
SVelter runtime

Steps	Runtime(s)	Process	Memory
Setup	594	Intel Xeon E7–4860	8 GB
Null model	885		8 GB
BPSearch	1300–10,000 per chromosome		8 GB
BPIntegrate	477		8 GB
SVPredict	150–5200 per chromosome		8 GB
SVIntegrate	258		8 GB

event, which for complex SVs may include more than two breakpoints and sometimes multiple chromosomes. The "ref" and "alt" columns describe the structure of the region in reference and sample genomes, with each letter stands for a genomic block defined by adjacent breakpoints, from left to right in alphabetical order. The last column, "other_info" provides additional information required to fully demonstrate the complex SVs, such as the insert point for a dispersed duplication.

3.8 Runtime

The software was run on a 2 Intel Xeon Intel Xeon E7–4860 processors with 4GB RAM per node, with processing time on 30X NA12878 whole genome sequencing listed by step in Table 6.

4 Notes

1. Only paired end sequences properly aligned and recorded in either bam for cram formats are accepted by SVelter for SV discovery.
2. During the setup step, absolute path for the reference sequences, working directory and SVelter support folder are required in this step.

References

1. Zarrei M, MacDonald JR, Merico D, Scherer SW (2015) A copy number variation map of the human genome. Nat Rev Genet 16(3):172–183
2. Mills RE, Walter K, Stewart C, Handsaker RE, Chen K, Alkan C, Abyzov A et al (2011) Mapping copy number variation by population-scale genome sequencing. Nature 470(7332): 59–65
3. Brand H, Pillalamarri V, Collins RL, Eggert S, O'Dushlaine C, Braaten EB, Stone MR et al (2014) Cryptic and complex chromosomal aberrations in early-onset neuropsychiatric disorders. Am J Hum Genet 95(4):454–461
4. Chiang C, Jacobsen JC, Ernst C, Hanscom C, Heilbut A, Blumenthal I, Mills RE et al (2012) Complex reorganization and predominant

non-homologous repair following chromosomal breakage in karyotypically balanced germline rearrangements and transgenic integration. Nat Genet 44(4):390–397

5. Stankiewicz P, Lupski JR (2010) Structural variation in the human genome and its role in disease. Annu Rev Med 61:437–455
6. Chen K, Wallis JW, McLellan MD, Larson DE, Kalicki JM, Pohl CS, McGrath SD et al (2009) BreakDancer: an algorithm for high-resolution mapping of genomic structural variation. Nat Methods 6(9):677–681
7. Ye K, Schulz MH, Long Q, Apweiler R, Ning Z (2009) Pindel: a pattern growth approach to detect break points of large deletions and medium sized insertions from paired-end short reads. Bioinformatics 25(21):2865–2871
8. Layer RM, Chiang C, Quinlan AR, Hall IM (2014) LUMPY: a probabilistic framework for structural variant discovery. Genome Biol 15(6):R84
9. Handsaker RE, Korn JM, Nemesh J, McCarroll SA (2011) Discovery and genotyping of genome structural polymorphism by sequencing on a population scale. Nat Genet 43(3):269–276
10. Rausch T, Zichner T, Schlattl A, Stütz AM, Benes V, Korbel JO (2012) DELLY: structural variant discovery by integrated paired-end and split-read analysis. Bioinformatics 28(18):i333–i339
11. Alkan C, Coe BP, Eichler EE (2011) Genome structural variation discovery and genotyping. Nat Rev Genet 12(5):363–376
12. Zhao X, Emery SB, Myers B, Kidd JM, Mills RE (2016) Resolving complex structural genomic rearrangements using a randomized approach. Genome Biol 17(1):126
13. Pendleton M, Sebra R, Pang AWC, Ummat A, Franzen O, Rausch T, Stütz AM et al (2015) Assembly and diploid architecture of an individual human genome via single-molecule technologies. Nat Methods 12(8):780–786
14. Parikh H, Mohiyuddin M, Lam HYK, Iyer H, Chen D, Pratt M, Bartha G et al (2016) Svclassify: a method to establish benchmark structural variant calls. BMC Genomics 17(January):64

Chapter 14

Analysis of Population-Genetic Properties of Copy Number Variations

Lingyang Xu, Liu Yang, Derek M. Bickhart, JunYa Li, and George E. Liu

Abstract

While single nucleotide polymorphisms (SNPs) are typically the variant of choice for population genetics, copy number variations (CNVs) which comprise insertions, deletions and duplications of genomic sequences, is also an informative type of genetic variation. CNVs have been shown to be both common in mammals and important for understanding the relationship between genotype and phenotype. Moreover, population-specific CNVs are candidate regions under selection and are potentially responsible for diverse phenotypes.

Key words Copy number variation, Genetic diversity, Population genetics, Positive selection

1 Introduction

Copy number variations (CNVs) are large-scale insertions and deletions, existing as one type of complex multiallelic variants within diverse populations [1, 2]. Compared to single nucleotide polymorphisms (SNPs), CNVs involve more genomic sequences and have potentially greater effects, including changing gene structure and dosage, altering gene regulation and exposing recessive alleles, and CNVs were shown to be important in both normal phenotypic variability and disease susceptibility [3, 4]. Investigating the population genetics and evolutionary selections of CNVs could enable us to understand their origins and impacts [5–8].

Population structure analyses based on human CNVs have revealed results largely consistent with those based on SNPs of similar number [9]. For instance, based on hybrid genotyping arrays, up to 90% of human CNVs can be revealed by integrated investigation of SNPs [10]. On the other hand, multiple lines of evidence also suggest CNVs could serve as an extra

The original version of this chapter was revised. A correction to this chapter can be found at https://doi.org/10.1007/978-1-4939-8666-8_17

Derek M. Bickhart (ed.), *Copy Number Variants: Methods and Protocols*, Methods in Molecular Biology, vol. 1833, https://doi.org/10.1007/978-1-4939-8666-8_14, © Springer Science+Business Media, LLC, part of Springer Nature 2018

genomic resource and provide important insights into the origins and substructure of populations [8, 11–14]. Additionally, population-specific CNVs are candidate regions under selection and potentially responsible for diverse phenotypes [3, 11, 15–17]. Population genetics analyses is often predicated on the biallelic state of marker SNPs—which can have three marker states—whereas identified CNVs can have multiple allelic variants. This represents a major impediment towards the use of highly refined, biallelic-state, mathematical models in the analysis of CNV genotypes for population surveys.

In this chapter, we illustrate serial steps for comprehensive population genetics study of CNVs by focusing on the diversity, population structure, and selection of identified CNVs within eight representative cattle breed. This Chapter offers example from our previous work [18], which investigated CNVs in cattle originating from European taurine, indicine, and African taurine breeds of the Bovine HapMap DNA panel [19]. Through this chart, we would like to guide readers the analysis steps of population-genetics study of CNVs, and we also provided several important steps of CNVs analysis including command line and R code to investigate population genetics of CNV in cattle.

2 Methods

2.1 Sample Selection

For this population genetic study, unrelated individuals were selected using genotype data based on Pi-hat-value (Pi-hat<0.25). Pi-hat value was estimated with genome option implemented in PLINK software. For instance, we have pop.ped and pop.map in PLINK format, the following command can be used to generate the table with estimated Pi-hat value for pairwise individuals.

```
plink --file pop --genome --out pop --cow --noweb
```

Then, unrelated individuals were kept for downstream analyses. Raw data was extracted from Illumina's GenomeStudio application (*see* **Note 1**).

2.2 CNVs Segmentation and Genotyping

To facilitate population genetic analysis, CNV segments were extracted from Golden Helix SVS 8.0 using the default multivariate option (*see* **Note 2**). After merging across all individuals, we discovered 263 nonredundant CNVs which are commonly shared within the whole population.

Since the SVS multivariate option was developed to identify moderate to high frequency CNVs, only segments with frequencies above 1% were retained for further analysis in order to filter away potential false positive calls. The detailed SVS multivariate option can be found in the Web (http://doc.goldenhelix.com/SVS/latest/svsmanual/copy_number_analysis.html).

Finally, a total of 257 CNVs were retained and used to categorize the samples as one of three types (loss, neutral and gain events) according to a three-state model with strict threshold levels of marker mean log R Ratio (LRR) ± 0.3. While all 257 CNVs were used for frequency and V_{ST} calculations, only the 184 deletion CNV regions were used in all other subsequent population genetic analyses (*see* **Note 3**).

We then utilized the copy number analysis module (CNAM) under the multivariate option to segment chromosomes with default set, and a significance level of $p = 0.01$ for pairwise permutations ($n = 1000$) as described previously. The three state covariates with a comparatively strict threshold (segment mean 0.3) was used to genotype the CNVs as one of three type (loss, neutral and gain events) across all the samples. It was noted that the multivariate method tends to detect the common deletions with relative small sizes across multiple samples.

2.3 Population Differences Across Population

We generated the mean LRR matrix from Golden Helix SVS, row presents individual, while the column presents CNV segment. The R function heatmap.2 (http://www.inside-r.org/packages/cran/gplots/docs/heatmap.2) was used to graph the segment mean LRR values and generate hierarchical cluster dendrograms using 257 CNVs for all animals.

```
Seg257<-read.table("Seg_257.
txt",header=T);dim(Seg257)
library(gplots)
#help(heatmap.2)
pdf("Seg_renames_257Seg_2.pdf")
seg257_heatmap <- heatmap.2(as.matrix(Segs257),
Rowv=NA, Colv=NA, col = redblue(256),
scale="column")
dev.off()
```

We first checked the normalized LRR distribution for all 184 deletion CNVs. The grand majority (99%) of deletions had two distinct peaks, representing neutral (around 0) and homozygous deletion (around −1) states, respectively. Only a couple of deletions had a few samples located in the midpoint between −1 and 0, suggesting a lack of heterozygous events. Additionally, it was difficult to define a universal threshold between homozygous or heterozygous deletions for all deletions, therefore, we decided to categorize all deletion events using one state: homozygous deletion. To use population genetic programs originally developed for SNPs, we manually recoded each 184 deletion CNVs by converting a loss event into "12" and a neutral event into "22," where "12" represented a homozygous deletion.

We then performed multidimensional scaling (MDS) and admixture analysis to determine how 205 unrelated individuals were clustered according to these CNV genotypes. Using a total of 184 deletion CNVs, we performed MDS analysis of pairwise genetic distance (4 dimensions) to detect the relationship between populations with PLINK 1.07 (-mds -plot 4). For a separate verification, we also carried out the cluster analysis based on mean LRR values using *prcomp* function in R.

Prepare the PLINK format file (PED and MAP) from the CNV genotypes information based on deletions. PLINK files: del_184.ped and del_184.map. Pairwise IBS metric was calculated and clustering of pairwise IBS metrics were evaluated based on 184 deletions using following command lines.

```
>plink --file del_184 --genome --out del_184 --cow
--noweb
>plink --file del_184 --read-genome ibs.genome
--cluster --ppc 1e-3 --mds-plot 2 --out del_184
--cow --noweb
```

Finally, we obtained del_184.mds file for MDS plot using R plot function.

To perform multidimensional scaling analysis on the N x N matrix of genome-wide IBS pairwise distances, use the --mds-plot option in conjunction with --cluster. This command takes a single parameter, the number of dimensions to be extracted. Plotting the C1 values against C2, for example, will generate a scatter plot in which each point represents an individual; the two axes correspond to a reduced representation of the data in two dimensions, which can be useful for identifying any clustering.

Genetic distance matrix was genetated using the command as follow

```
>plink --file del_184 --cluster --distance-
matrix --out del_184 --noweb --cow
```

This step can generate "del_184.mdist" file. To explore the admixture analysis using 184 deletions, we prepared the input file for STRUCTURE software using the recode option --recode-structure in PLINK.

```
> plink --file del_184 --recode-structure
--cow --noweb
```

Then, this command line generates the str file for STRUCTURE, here the second line of str file containing position information should be removed to fit special format in STRUCTURE "Date file stores dada for individuals in a single line". Population structure was examined using STRUCTURE 2.3. Each admixture analysis was performed using 5000 replicates and 2000 burn-in cycles under admixture and allele frequencies correlated models. The detailed usage of STUCTURE, refer to web site from Pritchard Lab, Stanford University http://web.stanford.edu/group/pritchardlab/structure_software/release_versions/v2.3.4/html/structure.html.

Neighbor-joining clustering analysis was performed using PHYLIP 3.69 (http://www.phylip.com/) based on pairwise genetic distance. Pairwise genetic distance (D) between individuals was calculated using PLINK 1.07, where D = 1-[IBS2 + 0.5IBS1)/N], and

IBS2 and IBS1 are the number of loci that share either 2 or 1 alleles identical by state (IBS), respectively and the N is the number of loci. The clustering dendrograms were plotted in Figtree 1.3.1 (http://tree.bio.ed.ac.uk/software/figtree/). Based on del_184.mdist, reformate the distance file for executable file neighbor.exe in PHYLIP package. Detailed usages of PHYLIP and its programs were presented at http://evolution.genetics.washington.edu/phylip.html. Next, executable file neighbor.exe was used to generate the outtree file and can be loaded into Figtree 1.3.1 software.

2.4 Signatures of Selection for CNVs or CNVR

To detect the lineage differentiated CNV events, we calculated V_{ST} for each CNV using the following equation: $(V_T - V_S)/V_T$, where V_T is the total variance in mean LRRs across all individuals and V_S is the average variance in cattle within each breed (*see* **Note 4**).

The test file contains mean LRRs of 257 segments across all individuals. The first column is population ID, and from second column to last were mean LRRs for 257 segments, the calculations of V_{ST} was showed as follows, and the map file (cnv-map) for each CNV segment was required, the head of map file includes first SNP name, chromosome, segment start, and segment end.

```
###R code for estimating the VST between pair-wise
groups.
###Read ped file
setwd("~://PATH/Vst)
CNV<-read.table("Segs257_min3_Max100_tran_1.txt",
header=F)
###Read map file
CNVmap<-read.table("cnvmap.txt", header=F)
###Calculate the pair-wise VST between BT and BI
###Extract population from Bos taurus group.
BTHOL<-CNV[CNV[,1]=="HOL",];dim(BTHOL)
BTANG <-CNV[CNV[,1]=="ANG",];dim(BTANG)
BTHFD <-CNV[CNV[,1]=="HFD",];dim(BTHFD)
BTBWS <-CNV[CNV[,1]=="BWS",];dim(BTBWS)
###Merge to Bos taurus (BT) group
BT<-rbind(BTHOL,BTANG, BTHFD, BTBWS);dim(BT);
nBT<-dim(BT)[1]
nBTHOL<-dim(BTHOL)[1]
nBTANG<-dim(BTANG)[1]
###Extract population Bos indican
BINEL <-CNV[CNV[,1]=="NEL",];dim(BINEL)
BIBRM <-CNV[CNV[,1]=="BRM",];dim(BIBRM)
###Merge to Bos indicine (BI) group
BI<-rbind(BINEL,BIBRM);dim(BI)
nBI<-dim(BI)[1]
############################
```

```
###Calculate variance in BT group
BTtype<-BT[,2:258];dim(BTtype)
SX2 <- function(x)[ var(x) ]
BTvar <- apply(BTtype,2,SX2)
### Calculate variance in BI group
BItype<-BI[,2:258]
SX2 <- function(x)[ var(x)]
BIvar <- apply(BItype,2,SX2)
###Calculate variance among all unrelated individu-
als
Atype<-rbind(BTtype,BItype)
ABTBIvar <- apply(Atype,2,SX2)
###Calculate weight mean of variance across two
groups (BT and BI)
wBTBIvar <- (BTvar*nBT+BIvar*nBI)/(nBT+nBI)
###Calculate VST for BT vs BI
VstBTBI<-(ABTBIvar-wBTBIvar)/ABTBIvar
length(VstBTBI)
###Combine the map information and VST values
ctVstBTBI<-cbind(CNVmap,VstBTBI)
###Assign the column names for the output file
names(ctVstBTBI)<-c("Chr","CNV","Start","Strand","c
tVstBTBI")
write.table(ctVstBTBI,"wctVstBTBI_205_
OK.txt",sep="\t",quote = FALSE)
```

3 Notes

1. The intensity data of SNP probes were generated using the Illumina BovineHD SNP array. After exporting the DSF file from GenomeStudio Software, we imported Log R Ratio (LRR) into Golden Helix SNP & Variation Suite (SVS) 8.0 (Golden Helix Inc., Bozeman, MT, USA) and then mapped onto the 29 autosomes of Bos taurus genome assembly UMD 3.1. The LRR was then normalized using the default GC correlation file to correct the waviness caused by the GC content.
2. This chapter provides population differences analysis based on common CNVs identified using multivariate option implemented in Golden Helix SNP & Variation Suite (SVS). However, most of CNV analysis pipeline may detect the CNV by individual, and generate the individual CNV calling, with different coordinate across population. Thus, in the other side, population differences analysis can also be conducted based on CNVR, individual CNV was merged into CNVR, and the

average log RR were estimated for each CNVR across population. Thus, the similar analysis may also be considered as previously indicated in this chapter, including hierarchical cluster and V_{ST} comparison based on CNVR.

3. CNV deletion can act as same as SNP due its simple genotype. Thus, to facility the downstream analysis, we conducted population genetics analysis by CNV deletion and CNV duplications separately, this both deletion and duplication were divided into three state genotype. For instance, CNV deletion can be coded as (0, 1, and 2), while CNV duplication can be coded as (3, 4, and 5). Using this conversion, some methods and tools for population genetic based on SNP can be utilized directly for CNV with three states genotype despite of some degree of bias may exist.
4. The detection of CNV has been extensively explored as the development of recent high-throughput technology including aCHG, SNP array, and NGS. (1) aCGH, whole-genome CGH arrays contain hundreds of oligonucleotide probes that were designed and fabricated on a single slide to provide an evenly distributed coverage using genome assemblies. The CNVs were represented by gains and losses of normalized fluorescence intensities relative to the reference. V_{ST} is calculated by considering $(V_T - V_S)/V_T$, where V_T is the variance in $\log_2$ ratios, and the $\log_2$ Ratios can be extracted from oligonucleotide probes [20]. (2) SNP, SNP arrays have been routinely used for human CNV detection and analysis. Compared to CGH arrays which only report relative signal intensities, SNP arrays collect normalized total intensities (Log R Ratio-LRR) and allelic intensity ratios (B allele frequency-BAF) which represent overall copy numbers and allelic contrasts. V_{ST} is calculated by considering $(V_T - V_S)/V_T$, where, V_T is the variance in $\log_2$ Ratios, and the $\log_2$ Ratios can be extracted from SNPs probes within CNVR across populations [18]. (3) NGS, the advent of next-generation sequencing (NGS) and complementary analysis programs has provided better approaches to systematically identify CNVs at a genome-wide level. These sequence-based approaches, which are becoming more popular due to the ongoing developments and cost decreases in NGS, allow CNV reconstruction at a higher effective resolution and sensitivity. Read depth (RD) methods were used to analyze NGS data to contribute high-resolution CNV calls with the capability of determining exact CN values for each genetic locus in an individual [21]. Population-specific CNVs were estimated using the statistic V_{ST} is calculated by considering $(V_T - V_S)/V_T$, where V_T is the variance in normalized copy numbers (estimated by read counts) among all unrelated individuals and V_S is the average variance within each population, weighted for population size.

Acknowledgments

This work was supported in part by Agricultural Science and Technology Innovation Program of China (ASTIP-IAS-TS-16). L.Y.X was supported by the Elite Youth Program in Chinese Academy of Agricultural Sciences.

References

1. Scherer SW, Lee C, Birney E et al (2007) Challenges and standards in integrating surveys of structural variation. Nat Genet 39(7 Suppl):S7–S15
2. Mills RE, Walter K, Stewart C et al (2011) Mapping copy number variation by population-scale genome sequencing. Nature 470(7332):59–65
3. Brown KH, Dobrinski KP, Lee AS et al (2012) Extensive genetic diversity and substructuring among zebrafish strains revealed through copy number variant analysis. Proc Natl Acad Sci U S A 109(2):529–534
4. Zhang F, Gu W, Hurles ME et al (2009) Copy number variation in human health, disease, and evolution. Annu Rev Genomics Hum Genet 10:451–481
5. Sudmant PH, Kitzman JO, Antonacci F et al (2010) Diversity of human copy number variation and multicopy genes. Science 330(6004):641–646
6. Conrad DF, Hurles ME (2007) The population genetics of structural variation. Nat Genet 39(7 Suppl):S30–S36
7. Sjodin P, Jakobsson M (2012) Population genetic nature of copy number variation. Methods Mol Biol 838:209–223
8. Narang A, Jha P, Kumar D et al (2014) Extensive copy number variations in admixed Indian population of African ancestry: potential involvement in adaptation. Genome Biol Evol 6(12):3171–3181
9. Jakobsson M, Scholz SW, Scheet P et al (2008) Genotype, haplotype and copy-number variation in worldwide human populations. Nature 451(7181):998–1003
10. McCarroll SA, Kuruvilla FG, Korn JM et al (2008) Integrated detection and population-genetic analysis of SNPs and copy number variation. Nat Genet 40(10):1166–1174
11. Berglund J, Nevalainen EM, Molin AM et al (2012) Novel origins of copy number variation in the dog genome. Genome Biol 13(8):R73
12. Redon R, Ishikawa S, Fitch KR et al (2006) Global variation in copy number in the human genome. Nature 444(7118):444–454
13. Kato M, Kawaguchi T, Ishikawa S et al (2010) Population-genetic nature of copy number variations in the human genome. Hum Mol Genet 19(5):761–773
14. Campbell CD, Sampas N, Tsalenko A et al (2011) Population-genetic properties of differentiated human copy-number polymorphisms. Am J Hum Genet 88(3):317–332
15. Iskow RC, Gokcumen O, Lee C (2012) Exploring the role of copy number variants in human adaptation. Trends Genet 28(6):245–257
16. Perry GH, Dominy NJ, Claw KG et al (2007) Diet and the evolution of human amylase gene copy number variation. Nat Genet 39(10):1256–1260
17. Xue Y, Sun D, Daly A et al (2008) Adaptive evolution of UGT2B17 copy-number variation. Am J Hum Genet 83(3):337–346
18. Xu L, Hou Y, Bickhart DM, Zhou Y et al (2016) Population-genetic properties of differentiated copy number variations in cattle. Sci Rep 6:23161
19. Bovine HapMap C, Gibbs RA, Taylor JF et al (2009) Genome-wide survey of SNP variation uncovers the genetic structure of cattle breeds. Science 324(5926):528–532
20. Liu GE, Hou YL, Zhu B et al (2010) Analysis of copy number variations among diverse cattle breeds. Genome Res 20(5):693–703
21. Bickhart DM, Xu L, Hutchison JL et al (2016) Diversity and population-genetic properties of copy number variations and multicopy genes in cattle. DNA Res 23(3):253–262

Chapter 15

Validation of Genomic Structural Variants Through Long Sequencing Technologies

Xuefang Zhao

Abstract

Although numerous algorithms have been developed to identify large chromosomal rearrangements (i.e., genomic structural variants, SVs), there remains a dearth of approaches to evaluate their results. This is significant, as the accurate identification of SVs is still an outstanding problem whereby no single algorithm has been shown to be able to achieve high sensitivity and specificity across different classes of SVs. The method introduced in this chapter, VaPoR, is specifically designed to evaluate the accuracy of SV predictions using third-generation long sequences. This method uses a recurrence approach and collects direct evidence from raw reads thus avoiding computationally costly whole genome assembly. This chapter would describe in detail as how to apply this tool onto different data types.

Key words VaPor, Long read sequencing, Structural Variants

1 Introduction

Structural variants (SVs) are one of the major forms of genetic variation in humans and have been revealed to play important roles in numerous diseases including cancers and neurological disorders [1, 2]. Various approaches have been developed and applied to paired-end sequencing to detect SVs in whole genomes [3–5]; however, individual algorithms often exhibit complementary strengths that sometimes lead to disagreements as to the precise structure of the underlying variant. The emergence of long read sequencing technology, PacBio or Oxford Nanopore [6–8], can deliver reads ranging from kilobase to megabase, thus providing direct evidence for the presence of an SV. Current strategies [6, 7, 9] that directly compare genome-wide assembled long contigs to reference genome to achieve minimized error rate, requires high read depth and significant computing power and are currently impracticable for many ongoing research studies.

The additional information obtained from using long reads can still be leveraged to improve variant calling, however. Indeed,

Derek M. Bickhart (ed.), *Copy Number Variants: Methods and Protocols*, Methods in Molecular Biology, vol. 1833,
https://doi.org/10.1007/978-1-4939-8666-8_15,

such approaches have already been implemented to combine high depth Illumina sequencing with lower depth PacBio reads to improve error correction and variant calling in the context of de novo genome assembly [10]. With structural variation, the current state of the art is to use long reads to manually assess potential SVs using subsequent recurrence (dot) plots [11], where the sequences are compared against the reference through a fixed size sliding window (k-mer) and the matches are plotted for visual inspection. The k-mer method is of higher robustness compared to direct sequences comparison [12], which is why these types of dot plots have been used for decades to examine the specific features of sequence alignments [13]. However, they require manual curation and, coupled with the computational costs of sequence assembly, are time-consuming and inefficient at scale for the high throughput validation of large sets of SVs.

The tool presented here, VaPoR, investigates and scores each SV by autonomously analyzing the recurrence of k-mers within a local read against both an unmodified reference sequence at that loci as well as a rearranged reference pertaining to the predicted SV structure. A positive score of each read on the altered reference, normalized against the score of the read on the original reference, supports the predicted structure. A baseline model is constructed as well by interrogating the reference sequence against itself at the query location. This approach can quickly and accurately distinguish true from false positive predictions of both simple and complex SVs as well as their underlying genotypes and is also able to assess the breakpoint accuracy of individual algorithms. The computational tool introduced in this chapter evaluates predicted genomic structural variants by adopting long read sequencing data. This method directly compares long sequences with a reference genome through straightforward computational metrics, thus achieving high efficiency in both run time and computing cost. In summary, this tool provides direct support for SVs estimated from indirect evidences, without high demand of sequencing or computing cost.

2 Materials

2.1 Data Preparation

We utilize sample NA12878, which has been examined extensively by various studies including HapMap, the 1000 Genomes Project (1KGP) and Genome in a Bottle (GIAB) using numerous microarray and sequencing platforms including both short and long read technologies. Its genomic SVs have been systematically discovered and represented in multiple publications [7, 14] and serves as a gold standard to illustrate the proper application of VaPoR. PacBio sequences introduced by Pendleton et al. are available:

` ` `

```
wget ftp://ftptrace.ncbi.nlm.nih.gov/giab/ftp/data/
NA12878/NA12878_PacBio_MtSinai/sorted_final_merged.
bam
```

To obtain corresponding reference genome:

```
wget
http://hgdownload.cse.ucsc.edu/goldenPath/hg19/big-
Zips/chromFa.tar.gz
```

For the purposes of this workflow, all datasets are reorganized with these linux commands:

```
mkdir workdir       #start a new folder named
'workdir'
mv sorted_final_merged.bam workdir/NA12878.PacBio.
bam
mv chromFa.tar.gz workdir/reference.fa.gz
gunzip workdir/reference.fa.gz
```

SV predictions are renamed as: workdir/NA12878.SVs.bed and workdir/NA12878.SVs.vcf
```

### 2.2 Software Preparation

Python (*https://www.python.org/*) and samtools (*http://samtools.sourceforge.net/*) are required for VaPoR. Required python packages can be installed autonomously through installation of vapor by following these commands:

```
git clone https://github.com/mills-lab/vapor.git
cd vapor
pip install  .
```

3 Methods

3.1 Evaluate SVs in Bed Format

Before applying VaPoR for any validation purpose, it is important to standardize the input variants into proper format. For input in *bed* format, a fourth column indicating SV type is required and should be preprocessed by the user. Here is an example of the input:

```
chr1    831220      833732      DUP
chr1    2093108     2095944     DEL
chr1    2994972     2995286     INV
chr12   28226407    28226407    INS_sequences
chr1    3935209     3935209     INS_191
```

For proper insertion validation, the inserted sequences should be described in the fourth column starting with "INS_". If sequence not available, approximate length of the inserted sequences should be specified instead.
```

To run vapor on NA12878 with SVs in bed format:

```
vapor bed \
--pacbio-input workdir/NA12878.PacBio.bam \
--reference workdir/reference.fa \
--sv-input workdir/NA12878.SVs.bed \
--output-path workdir/
```

3.2 Evaluate SVs in vcf Format

SVs kept in *vcf* format should at least keep these features in the info column: END, SVTYPE, SVLEN.

To run vapor on NA12878 with SVs in *vcf* format:

```
vapor vcf \
--pacbio-input workdir /NA12878.PacBio.bam \
--reference workdir/reference.fa \
--sv-input workdir /NA12878.SVs.vcf \
--output-path workdir/
```

3.3 Optional Parameters

To derive reliable scores, a minimum of three long reads are required by default to cover the predicted SVs by default. However, in situations where samples have been sequenced at low throughput (e.g., <5×) or assembled long contigs (typically 1–2× assembly coverage) were adopted for validation, this number can be adjusted with the "--PB-supp" parameter.

3.4 Output Format

VaPoR output is reported under "--output-path" with file name: SVinput.vapor.

Here is an example of the SVinput.bed.vapor:

```
CHR      POS       END       SVTYPE   VaPoR_qs VaPoR_gs
VaPoR_GT VaPoR_GQ VaPoR_Rec
chr1      26640516 26647387 INV
0.261932979463     1.0      1/1      1.14712576604
0.26,0.26,0.26,0.26
chr1      43592391 43595367 INV      0         0.0
0/0      1.14712576604      -5.37,-4.92,-4.92,-4.92
chr1      80327754 80330649 INV      0         0.0
0/0      1.14712576604      -18.49,-18.34,-17.43,-18.19
```

5 new columns representing VaPoR evaluation are added to each SV record:

VaPoR_QS: quality score of vapor evaluation

VaPoR_GS: genotype score of vapor evaluation,

VaPoR_GT: predicted genotype of evaluated SV

VaPoR_GQ: quality of predicted genotype

VaPoR_Rec: evaluation score of each long read adopted for evaluation; positive score supports the prediction, otherwise negative score (*see* **Note 1**).

For SVs in vcf format, evaluation will be added to the info column with the same key names as described above. An example of this output is given below:

```
chr1    2262045 SVID1      A       <DEL>  0       PASS
END=2262115;SVLEN=-70;SVTYPE=DEL; VaPor_
GS=0.0;VaPor_GT=0/0;VaPor_GQ=1.15;VaPor_REC=-
0.04,0.0,-0.04,-0.12;VaPor_GS=0.0;VaPor_
GT=0/0;VaPor_GQ=1.15;VaPor_REC=-0.04,0.0,-0.04,-
0.12
chr1    2281172 SVID12      C
<DEL>  82.4    PASS    END=2282431;SVLEN=-
1259;SVTYPE=DEL; VaPor_GS=NA;VaPor_GT=NA;VaPor_
GQ=NA;VaPor_REC=NA;VaPor_GS=NA;VaPor_GT=NA;VaPor_
GQ=NA;VaPor_REC=NA
```

4 Note

1. Please note that multiallelic copy number variants and complex SVs cannot be properly evaluated by the current version of VaPoR, so that no vapor information would be added to those SVs. "NA" from VaPoR implies two situations: the predicted SV is not covered by enough reads for an accurate assessment, or the SV resides in a highly repetitive region. Reliable predictions can be rarely obtained from either situation.

References

1. Alkan C, Coe BP, Eichler EE (2011) Genome structural variation discovery and genotyping. Nat Rev Genet 12(5):363–376
2. Brand H et al (2014) Cryptic and complex chromosomal aberrations in early-onset neuropsychiatric disorders. Am J Hum Genet 95(4):454–461
3. Layer RM et al (2014) LUMPY: a probabilistic framework for structural variant discovery. Genome Biol 15(6):R84
4. Rausch T et al (2012) DELLY: structural variant discovery by integrated paired-end and split-read analysis. Bioinformatics 28(18):i333–i339
5. Zhao X et al (2016) Resolving complex structural genomic rearrangements using a randomized approach. Genome Biol 17(1):126
6. Chaisson MJP et al (2014) Resolving the complexity of the human genome using single-molecule sequencing. Nature 517 (7536):608–611
7. Pendleton M et al (2015) Assembly and diploid architecture of an individual human genome via single-molecule technologies. Nat Methods 12(8):780–786
8. Rhoads A, Au KF (2015) PacBio sequencing and its applications. Genomics Proteomics Bioinformatics 13(5):278–289

9. Shi L et al (2016) Long-read sequencing and de novo assembly of a Chinese genome. Nat Commun 7:12065
10. Koren S et al (2012) Hybrid error correction and de novo assembly of single-molecule sequencing reads. Nat Biotechnol 30(7):693–700
11. Huddleston J et al (2016) Discovery and genotyping of structural variation from long-read haploid genome sequence data. Genome Res 27(5):677–685. https://doi.org/10.1101/gr.214007.116
12. Carvalho AB, Dupim EG, Goldstein G (2016) Improved assembly of noisy long reads by k-mer validation. Genome Res 26(12):1710–1720
13. Gibbs AJ, McIntyre GA (1970) The diagram, a method for comparing sequences. Its use with amino acid and nucleotide sequences. Eur J Biochem 16(1):1–11
14. Parikh H et al (2016) svclassify: a method to establish benchmark structural variant calls. BMC Genomics 17:64

Chapter 16

Structural Variation Detection and Analysis Using Bionano Optical Mapping

Saki Chan, Ernest Lam, Michael Saghbini, Sven Bocklandt, Alex Hastie, Han Cao, Erik Holmlin, and Mark Borodkin

Abstract

The need to accurately identify the complete structural variation profile of genomes is becoming increasingly evident. In contrast to reference-based methods like sequencing or comparative methods like aCGH, optical mapping is a de novo assembly-based method that enables better realization of true genomic structure. It allows for independently detecting balanced and unbalanced structural variants (SVs) from separate alleles and for discovering de novo events. Here we show how Bionano Genome Mapping creates de novo assemblies from native and intact, megabase-scale DNA molecules and uses those assemblies to detect a wide range of structural variants.

Key words Structural Variation, Insertion, Deletion, Inversion, Translocation, Long reads, Genomics, Genome structure, Optical mapping

1 Introduction

Sequence variation contributes to phenotypic variation among individuals. Structural variants (SVs) represent a major class of sequence variation that collectively accounts for a greater number of variable bases than single-nucleotide variants (SNVs). Recent analysis from the 1000 Genomes Project showed that human genomes contain a median of 18.4 Mbp of SVs [1]. SVs have been shown to impact gene expression [2] and are associated with human diseases such as autism [3] and cancer [4]. In plants and animals, the heritability of many complex phenotypes is the result of large structural variation (SVs) rather than SNPs. Tandem gene repeats and transposable element copy numbers play a role in a variety of phenotypes in maize [5]. For these reasons, comprehensive SV discovery and analysis is critical to understanding how

Saki Chan, Ernest Lam, and Michael Saghbini authors contributed equally to this work.

Derek M. Bickhart (ed.), *Copy Number Variants: Methods and Protocols*, Methods in Molecular Biology, vol. 1833, https://doi.org/10.1007/978-1-4939-8666-8_16,

genomic differences give rise to phenotypic differences among individuals.

Short-read sequencing-based variant analysis provides near-complete sensitivity for detecting SNVs in euchromatic regions of the genome; however, SV analysis using short reads remains challenging. Short reads may be used for alignment-based and assembly-based methods to detect SVs [6, 7]. Alignment-based methods rely on accurate read alignment. However, in repetitive and low-complexity regions in the genome, misalignment can lead to false positive SV calls. Assembly-based methods require either global or local assembly of reads as a first step. Short reads lack long-range information necessary to decipher structurally complex regions of the genome, resulting in misassembly or fragmented assembly. Also, homologous alleles are often collapsed incorrectly, resulting in haploid assemblies that represent only one allele or a chimeric mixture of both alleles. Regardless of the approach, one is limited to incomplete structural information.

1.1 Beyond Standard Short Reads

Alternative sample preparation methods allow for preservation of some long-range information. For example, a recent study using long-insert whole genome sequencing detected large and complex SVs [3] and concluded that their prevalence and complexity had been underestimated. Similarly, barcode linked reads derived from high molecular weight molecules enabled better resolution of complex rearrangements in cancer samples [8]. However, these approaches are still based on short reads and require inference of the underlying genome structure.

Long reads in the kilo-base pair range make assembly and SV analysis more accessible. In recent studies [9], the authors showed that the majority of SVs detected using long-read data were novel. While promising, these platforms have difficulties calling heterozygous SVs, their single-base error rates are high, and per sample costs and run times are prohibitive for population-scale studies. In addition, platform-specific challenges such as handling of homopolymer stretches for nanopore sequencing need to be improved to minimize the number of spurious SVs.

1.2 Next-Generation Optical Mapping

Optical mapping was initially introduced in the 1990s [10]; however, it was only recently that new developments were made to make it tractable for analysis of large genomes. Optical maps have always provided long-range information, by imaging whole intact single molecules of DNA in their native state. The recent innovation was to electrophorese the molecules into NanoChannels etched onto silicon wafers, in order to uniformly stretch the DNA (Fig. 1), so that measuring the distances between fluorescently labeled probes became simpler and much more accurate. Also, the NanoChannel array technology enabled massive parallel processing, boosting the throughput and reducing the cost per sample. Most importantly, the long range information preserved in native single molecules enabled effective de novo assembly of large portions of

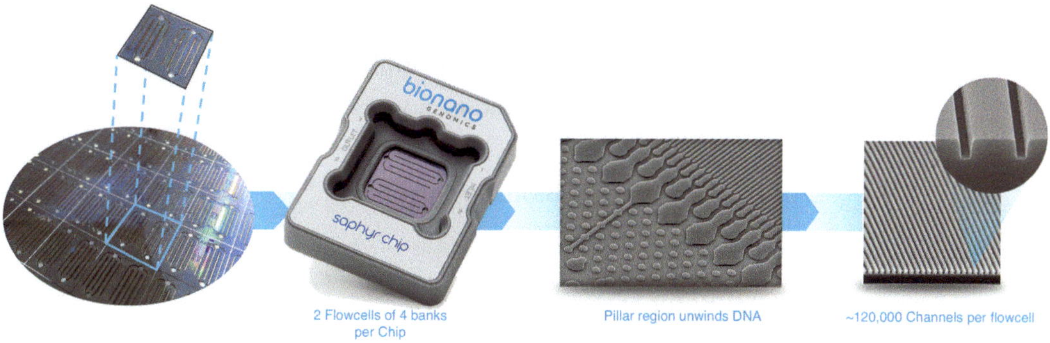

Fig. 1 Bionano Saphyr chips are made from silicon wafers, onto which millions of NanoChannels are etched

each haplotype, with proper linkages. With separate alleles assembled, there is visibility of SVs from both parents and hence zygosity information. This simplifies family-based inheritance studies. The mapping process is broken down into sample preparation, data generation and structural variant (SV) analysis.

2 Materials

1. Agarose (Bio-Rad).
2. Agarose Plug Molds (Bio-Rad).
3. Animal Tissue DNA Isolation Kit (Bionano Genomics).
4. Blood and Cell Culture DNA Isolation Kit (Bionano Genomics).
5. Plant DNA Isolation Kit (Bionano Genomics).
6. RBC Lysis Solution (Qiagen).
7. Ficoll-Paque PLUS (Fisher Scientific).
8. Cell Buffer (Bionano Genomics).
9. QiAamp DNA Blood Mini Kit (Qiagen).
10. 2% Formaldehyde Solution.
11. Ethanol.
12. Proteinase K (Qiagen).
13. Lysis Buffer (Bionano Genomics).
14. Screened Cap for 50 mL conical tube (Bio-Rad).
15. RNase A Solution (Qiagen).
16. Wash Buffer (Bionano Genomics).
17. Agarase (0.5 U/μL) (Thermofisher).
18. Dialysis Membrane (Millipore).
19. NLRS DNA Labeling Kit (Bionano Genomics).
20. DLS DNA Labeling Kit (Bionano Genomics).
21. Nickase (NEB).

22. Labeling Mix (Bionano Genomics).
23. Repair Mix (Bionano Genomics).
24. DNA Stain (Bionano Genomics).
25. DTT.
26. Flow Buffer (Bionano Genomics).

2.1 Equipment

1. HemoCue.
2. Rotor Stator (Qiagen).
3. Thermomixer.
4. Bionano Saphyr (Bionano Genomics).

3 Methods

3.1 Sample Preparation

Since optical mapping relies on long DNA molecules, short-read sequencing purification methods are not suitable for DNA isolation. Bionano Genomics adapted the plug lysis protocol used to construct BAC libraries for optical mapping. Briefly, cells/nuclei are embedded into an agarose matrix to protect DNA from mechanical shearing during the purification process. Agarose is then melted and solubilized, and the resulting ultra-high molecular weight (UHMW) DNA is further cleaned by drop dialysis prior to labeling at sequence-specific sites by nick translation.

3.1.1 Ultra-High Molecular Weight DNA Isolation

The first step in UHMW DNA isolation is the preparation of high quality cell/nuclei for embedding into a porous agarose matrix. This is important because the bulk of DNA purification is carried out by proteinase K digestion in the presence of detergents and heat followed by a series of washes to remove contaminants by means of diffusion while the DNA is still embedded in a plug. The final dialysis step after plug solubilization removes trace contaminants to ensure consistent labeling across samples. Embedding in low-melting point agarose plugs is done using agarose and plug molds from Bio-Rad. Cells/nuclei preparation and DNA purification are carried out using the appropriate Bionano Genomics kits—*see* Subheading "Preparation of Cells/Nuclei" below.

Preparation of Cells/Nuclei

A defined uniform suspension of cells/nuclei ensures high quality UHMW DNA at the proper concentration for effective labeling. For some samples such as cultured cell lines, isolated WBCs or PBMCs, and buccal washes, further preparation other than cell counting is not necessary. For samples such as tissues, the mass must be dissociated/homogenized before proceeding. Bionano Genomics offers specific kits for processing animal tissue and blood for subsequent plug lysis purification (Animal Tissue DNA Isolation Kit, Blood and Cell Culture DNA Isolation Kit). A kit is also available for processing plant samples (Plant DNA Isolation Kit).

For high quality WBC isolation, 3 mL of fresh blood is subjected to two rounds of differential lysis using 9 mL of RBC lysis solution. High quality PBMC isolation can be obtained using Ficoll-Paque PLUS. The resulting WBC/PBMC pellets are resuspended in Cell Suspension Buffer and quantitated for proper plug lysis targeting, by either counting with a HemoCue or by estimating DNA content using QiAamp DNA blood mini kit. For optimal plug lysis purification of blood cells, ~1 million cells if counting, or ~6 μg if estimating DNA content are targeted per plug.

For animal tissue processing, a defined tissue input is homogenized with a rotor stator after a brief fixation in 2% formaldehyde for fibrous tissue such a lung, or Dounce homogenized for soft tissue such as liver followed by fixation with an equal volume of alcohol. Cell/nuclei suspension corresponding to ~5 mg of tissue is targeted per plug. Low-density tissue may require a higher concentration possibly impacting DNA quality. For optimal plug lysis results, 3–8 μg calculated DNA yield from a tissue homogenate are targeted per plug. Input titration is useful if working with a new tissue for the first time to achieve quality DNA at the proper concentration.

Embedding of Cells/Nuclei in Agarose

The appropriate amount of cells/nuclei is resuspended in Cell Suspension Buffer to a final volume of 65 μL per plug to be made and equilibrated at 43 °C for up to 10 min if derived from cultured cells, WBCs or PBMCs, or kept at room temperature if derived from tissue. The cells/nuclei are mixed with 43 °C low melting-point agarose solution at 36 μL agarose per 65 μL cells/nuclei prior to casting plugs. Mixing should be done with the pipette volume set to 90% of the mixture volume, in order to prevent air bubble formation, which is detrimental to the structural integrity of the plug. Plugs are solidified at 4 °C for 15 min.

Plug Lysis Purification

Up to 5 plugs of the same cell/nuclei input are transferred to a 50 mL conical tube containing lysis solution prepared by adding Proteinase K enzyme to 2.5 mL Lysis Buffer. 167 μL of proteinase K is used for cultured cells and WBCs/PBMCs, while 200 μL is used for plant and animal tissue cells/nuclei. The conical tube is incubated on a 30 °C Thermomixer with intermittent mixing (10 s at 450 rpm followed by 10 min at 0 rpm) for 2 h. Fresh lysis solution is prepared, old lysis solution is removed, and the new lysis solution is added. To do this, a screened cap is used to drain the old solution, the tube is tapped on the benchtop to return plugs to the bottom, the screened cap is removed, and new lysis solution is added. The conical tube is then returned to the Thermomixer and incubated overnight. The order of the 2-h and overnight incubation steps can be reversed.

After Proteinase K treatment, the Thermomixer is set to 37 °C, 50 μL of RNase A solution is added, and the conical tube is incubated on the Thermomixer for 1 h. This step is not needed if working with WBCs/PBMCs.

The plugs are rinsed three times with 10 mL Wash Buffer followed by four washes in the same buffer. While washing, the conical tube is placed on an orbital shaker for 10 min at 180 rpm. The last wash should not be discarded if storing plugs for DNA recovery at a later date as plugs are stable for several weeks at 4 °C.

DNA Recovery

To recover UHMW DNA, plugs are washed five times with 10 mL of TE per wash. While washing, the conical tube is placed on an orbital shaker for 10 min at 180 rpm. One plug at a time is removed using a metal spatula, and excess buffer is removed by touching the corner of a tissue wipe to the spatula, taking care to not make direct contact with the plug. The plug is transferred to a 1.5 mL microcentrifuge tube and pulse spun to move the plug to the bottom of the tube. The tube is incubated at 70 °C for 2 min and then immediately moved to 43 °C for 5 min. 2 μL of Agarase (0.5 U/μL) is added and stirred gently with a pipette tip for 10 s. The tube is incubated at 43 °C for 45 min to digest the agarose.

A wide bore tip is used to transfer the solubilized plug to the center of a 0.1 μm dialysis membrane floated on 15 mL of TE buffer in a covered 6 cm Petri dish. The DNA droplet is left to dialyze for 45 min at room temperature. After dialysis, DNA is transferred to a new 1.5 mL microcentrifuge tube, using a wide bore tip and left to homogenize at room temperature overnight, after which it is stored at 4 °C. If the DNA is viscous after drop dialysis, pipet-mixing up to nine times with a narrow bore tip until the entire sample is taken up in a continuous flow can improve homogeneity during the overnight incubation.

3.1.2 QC: Evaluation of Isolated DNA

DNA is evaluated on clarity, homogeneity, viscosity, and concentration. Ideal DNA for optical mapping is clear, homogeneous, viscous, and at a concentration of 40–100 ng/μL. DNA that does not exhibit these qualities may be useable but may not produce optimal results. Most importantly, if the DNA is not viscous at all, it is likely fragmented and/or of too low concentration to use for labeling.

3.1.3 Fluorescent Labeling of DNA

Plug lysis recovered DNA is labeled using either the Bionano Prep™ DLS or NLRS DNA labeling Kit. Using the Direct Label and Stain (DLS) protocol, DNA is incubated with the DLE-1 enzyme in the presence of DL-Green. The enzyme is inactivated using proteinase K and excessive dye removed [11]. Alternatively, using Nick, Label, Repair, Stain (NLRS), single-stranded cuts are made on the DNA using nickases (site-specific restriction enzymes that cut one strand of DNA). Next, nick translation is performed with Taq Polymerase in the presence of fluorescently tagged nucleotides to label the nicks. Finally, nicks are repaired with Taq Ligase to restore molecule integrity.

Nicking

300 ng of DNA is combined with the appropriate nickase, buffer, and water to a final volume of 10 μL. The mixture is incubated at 37 °C for 2 h.

Labeling/Nick Translation

5 μL of Labeling Mix containing fluorescent nucleotides and Taq Polymerase is added and the mixture is incubated at 72 °C for 1 h.

Repair/Ligation

5 μL of Repair Mix containing Taq ligase with its cofactor is added and incubated at 37 °C for 30 min.

Staining

After fluorescent labeling and repair, DNA stain, DTT, water, and Flow Buffer are added to a final volume of 60 μL per 300 ng starting DNA, and mixed with a P200 and a wide bore tip. The mixture is left to rest overnight at room temperature, after which it is ready to run on a Bionano chip.

3.2 Data Generation

3.2.1 Using the Saphyr Instrument

After overnight rest, 8.5 μL of the sample is carefully loaded in the inlets of each Saphyr Chip™ flow cell and 120 s later into the outlets. The Saphyr clip is added onto the loaded chip to provide a hermetic seal, to protect the sample from contamination and evaporation. Electrodes on the clip dip in each of the inlets and outlets of the flow cells, and provide the contacts for the electrophoretic current. The chip is placed in the Saphyr instrument, the bundle arm lowered and the sample door closed.

Using Bionano Access to Start Runs and Monitor Live Performance Stats

Bionano Access™, a web-based software tool, provides a user-friendly interface for managing projects, analyzing run results, and providing data visualization.

Once experimental and run parameters are defined in Access, a run can be initiated. The Saphyr™ instrument will initiate electrophoresis. The instrument uses machine learning at the start and throughout the run to provide adaptive loading of DNA, optimize run conditions and maximize throughput.

When molecules are fully loaded into the NanoChannels of one flow cell, electrophoresis is halted and the entire surface of the NanoChannel array of that flow cell is rapidly imaged. During the imaging phase of the run, electrophoresis in the second flow cell is initiated. Cycles of loading of the NanoChannels followed by imaging are performed until sufficient data is collected.

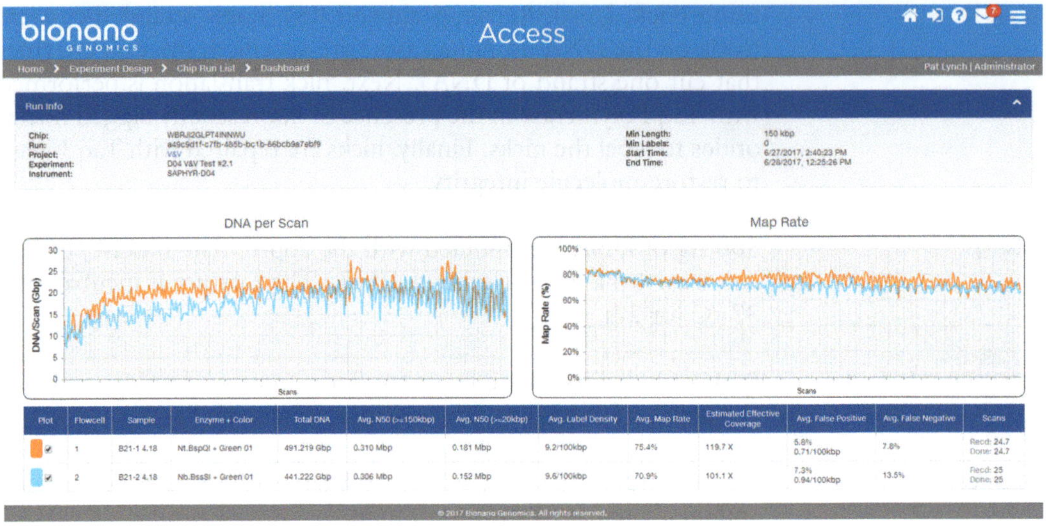

Fig. 2 Bionano Access dashboard view for monitoring run progress

Access provides a dashboard for an overview of a run (Fig. 2). Users can monitor the progress of the run and view real-time data metrics such as amount of data collected and alignment rate.

When human genomes are mapped, the instrument can automatically launch assembly and SV calling when sufficient data has been collected.

3.2.2 Conversion of Images into Digitized Molecules

Bionano Image Detect software creates digitized molecules from raw image data. Individual DNA molecules are identified, and the backbone stain signal of the DNA molecules is used to determine molecule position and size. The distance between the labels on each molecule is recorded to generate digitized molecules. The molecule data is collected into a BNX file, which provides the input for the de novo assembly.

Images generated by Saphyr are sent to the analysis server for real time data extraction during the run. Image detection is typically completed shortly after the run is finished.

3.3 Structural Variant Analysis

3.3.1 Automated Assembly and SV Calling Pipeline

A standard automated pipeline for de novo assembly and SV calling was developed by Bionano to enable comprehensive SV analysis. The Python-based pipeline manages job submission, drives execution of alignment and assembly tools, and provides data summary information. It features a haplotype-aware assembler designed to detect and differentiate parental alleles (Fig. 3). The de novo assembly algorithm is a custom implementation of the overlap–layout–consensus strategy. The assembler assembles digitized molecules from raw image data, and the final consensus maps are used as input for SV calling.

SV calls are made based on analyses of a multiple local alignment between consensus maps and the reference (Fig. 4).

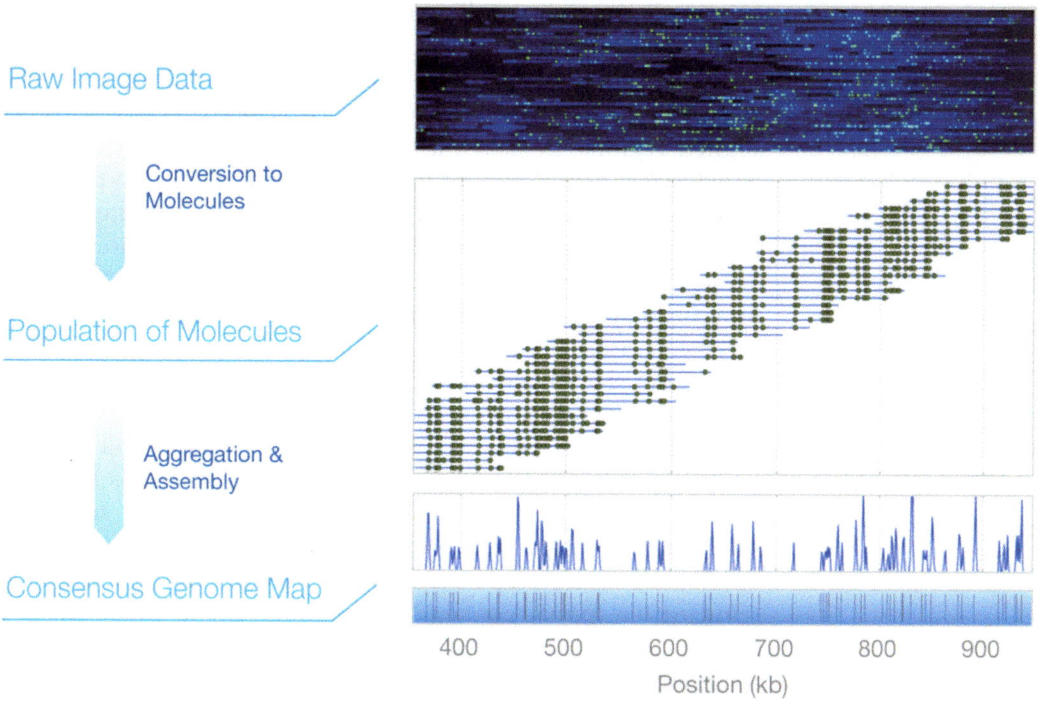

Fig. 3 Haplotype-aware assembler uses molecules from a digitized raw image to create a de novo assembly

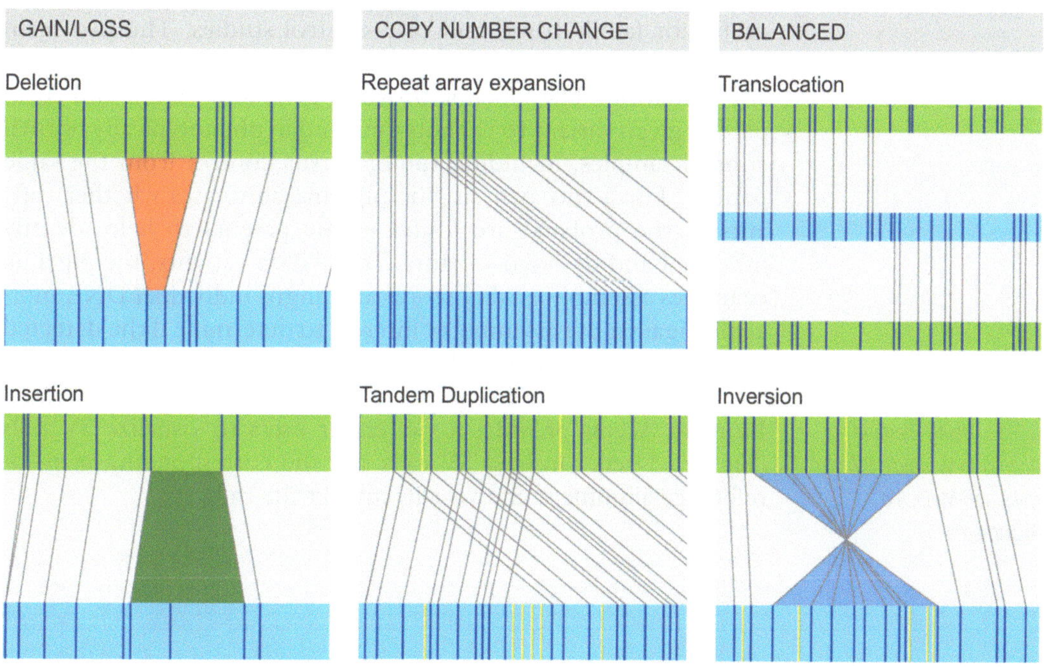

Fig. 4 Alignments of the consensus maps of each sample (blue) against the reference (green) shows the various types of structural variants identified

The pipeline supports calling of major SV types: insertions, deletions, inversions, and translocation breakpoints. Bionano Access also supports visualization and confidence-based filtering of these SV types. Poorly aligned or unaligned regions flanked by well-aligned regions are called as deletions or insertions, depending on whether there is gain or loss of sequence relative to the reference. Junctions of neighboring alignments with opposite orientations are identified as inversion breakpoints. Fusion points between distant regions of the genome are identified as translocation breakpoints. Intrachromosomal translocation breakpoints involve regions on the same chromosome but at least 5 Mbp away from each other. Interchromosomal translocation breakpoints involve regions on different chromosomes.

Zygosity and confidence are assigned to each SV call to facilitate downstream analysis. An SV call can be labeled as homozygous, heterozygous, or unknown. Confidence scores are scaled such that they range from 0 to 1.

SV calls can be exported in a dbVar compliant VCF file. This standard file format contains all genomic variants identified in a sample including SNVs, small indels, and SVs of various sizes. The VCF file generated by Bionano Access can be used in downstream analysis using a variety of existing tools.

3.3.2 Automated Variant Annotation Pipeline

Bionano Genomics developed a variant annotation pipeline (VAP) to help prioritize variants and to determine if a variant is relevant to the disease or phenotype of interest. In particular, it is useful for family-based and case–control studies. The two main components of the VAP are (1) variant annotation, and (2) variant validation. The pipeline provides gene annotation and compares a given variant to variants detected in phenotypically normal control samples, including tumor versus control from the same patient. For a trio analysis, the pipeline annotates whether variants in the proband are found in the parents to help identify inherited and de novo variants. To validate variants, the pipeline examines assembly quality scores and aligns individual DNA molecules against the assembly of interest to determine if the detected variants are well supported.

3.3.3 Using Bionano Access to Visualize SV Calls and Molecule Alignments

Bionano Access provides a viewer for users to visualize SV calls (Fig. 5). Users can load SV calls and the corresponding map-to-reference alignment and visually inspect the calls.

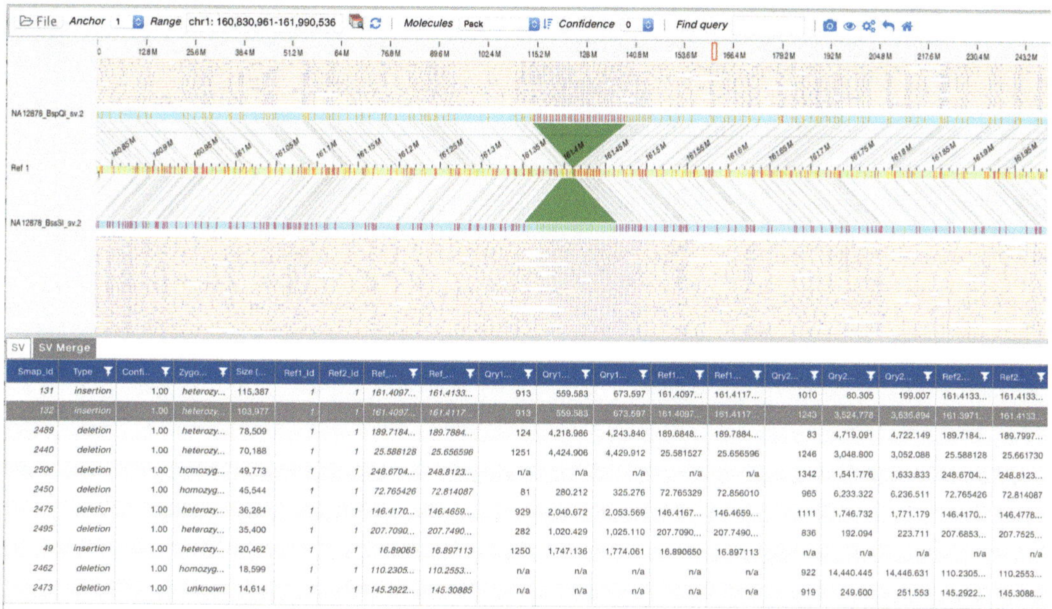

Fig. 5 Bionano Access structural variation (SV) visualization view for analysis of SV calls and the underlying alignment. The sample shows a repeat expansion relative to the reference hg19 (green) identified in two independent Bionano assemblies, labeled with Nt.BspQI (above reference) and with Nb.BssSI (below reference). Single molecule support for the consensus maps (blue) is also shown. The SV table (bottom of figure) enables easy navigation of the SV calls

References

1. Sudmant PH, Rausch T, Gardner EJ, Handsaker RE et al (2016) An integrated map of structural variation in 2504 human genomes. Nature 526:75–81
2. Chiang C, Scott AJ, Davis JR et al (2017) The impact of structural variation on human gene expression. Nat Genet 49:692–699
3. Collins RL, Brand H, Redin CE et al (2017) Defining the diverse spectrum of inversions, complex structural variation, and chromothripsis in the morbid human genome. Genome Biol 18:36
4. Zack TI, Schumacher SE, Carter SL et al (2013) Pan-cancer patterns of somatic copy number alteration. Nat Genet 45:1134–1140
5. Jiao Y, Peluso P, Shi J et al (2017) Improved maize reference genome with single-molecule technologies. Nature 546:524–527
6. Alkan C, Coe BP, Eichler EE (2011) Genome structural variation discovery and genotyping. Nat Rev Genet 12:363–376
7. Escaramís G, Docampo E, Rabionet R (2015) A decade of structural variants: description, history and methods to detect structural variation. Brief Funct Genomics 14:305–314
8. Greer SU, Nadauld LD, Lau BT et al (2017) Linked read sequencing resolves complex genomic rearrangements in gastric cancer metastases. Genome Med 9:57
9. Huddleston J, Eichler EE (2016) An incomplete understanding of human genetic variation. Genetics 202:1251–1254
10. Schwartz DC, Li X, Hernandez LI et al (1993) Ordered restriction maps of Saccharomyces cerevisiae chromosomes constructed by optical mapping. Science 262:110–114
11. Deschamps S, Zhang Y, Llaca V et al (2018) A chromosome-scale assembly of the sorghum genome using nanopore sequencing and optical mapping. bioRxiv 327817

Correction to: Analysis of Population-Genetic Properties of Copy Number Variations

Lingyang Xu, Liu Yang, Derek M. Bickhart, JunYa Li, and George E. Liu

Correction to
Chapter 14 in: Derek M. Bickhart (ed.), *Copy Number Variants: Methods and Protocols*, Methods in Molecular Biology, vol. 1833, https://doi.org/10.1007/978-1-4939-8666-8_14

The original version of the book was inadvertently published with incorrect spelling of the author name "Liu Yang". The author name has now been corrected.

The updated online version of this chapter can be found at
https://doi.org/10.1007/978-1-4939-8666-8_14
https://doi.org/10.1007/978-1-4939-8666-8

Derek M. Bickhart (ed.), *Copy Number Variants: Methods and Protocols*, Methods in Molecular Biology, vol. 1833,
https://doi.org/10.1007/978-1-4939-8666-8_17, © Springer Science+Business Media, LLC, part of Springer Nature 2018

Correction to: Analyses of Population-Genetic Properties

Index

Derek M. Bickhart (ed.), *Copy Number Variants: Methods and Protocols*, Methods in Molecular Biology, vol. 1833, https://doi.org/10.1007/978-1-4939-8666-8, © Springer Science+Business Media, LLC, part of Springer Nature 2018

MIX
Papier aus verantwortungsvollen Quellen
Paper from responsible sources
FSC® C105338

If you have any concerns about our products,
you can contact us on
ProductSafety@springernature.com

In case Publisher is established outside the EU,
the EU authorized representative is:
Springer Nature Customer Service Center GmbH
Europaplatz 3, 69115 Heidelberg, Germany

Printed by Libri Plureos GmbH
in Hamburg, Germany